# THE DIGITAL DISRUPTION

The Past, Present and Future of Digitalization and Its
Impact on The World We Live In

SECOND EDITION

## OSAMA MOHAMMED ALZOUBI

# DEDICATION

All I have, and everything I have achieved are due to the
tremendous effort, dedication, and sacrifices of my great
parents, Mohammed & Sara. Not only do I owe all of
that to them, but I also owe my life to them. The least I
can do is dedicate this book to them and to their effort.
My parents have always been an inspiration, support,
and guidance throughout my entire life, and also through
my professional career. My father has passed away, may
ALLAH bless his soul, but I know, that if he were to see
or read this, he would be very proud and happy. I am
very blessed to have my mother, and the first copy will
be printed, signed & gifted to her in her honor, may
ALLAH give her long life and the best of health. I also
dedicate this book to my wonderful wife and marvelous
children for their endless support and fabulous
encouragement throughout the journey. My parents are
in my heart, and my family are my eyes. I will never
forget how much I owe them.

# COPYRIGHT

First Publication, **September 2019**

ISBN:           987-0-578-57145-4

Email:          support@thedigitaldisruption.net

Website:        www.thedigitaldisruption.net

# TABLE OF CONTENTS

# INTRODUCTION

Looking at the current state of the world in recent times, we can boldly say that technology has evolved over the years, while having a profound impact on every aspect of our lives for several decades. However, the rate at which technological transformations are occurring has become faster over the years and is beginning to have an enormously disruptive effect on the global culture of economy, business, innovations, governance, and livelihood. And this disruption is predictively expected to intensify over the next coming decades.

Organizations now make use of technology to create new business models, moving closer to meeting the conscious awareness of the market's needs, while improving innovations and productivity at the same time. However, this increases the complexity of the systems and the difficulty of keeping up with the required skill and increased need for interdependence. These challenges are exacerbated by a rapid change in global market expectations and workforce culture. While executives consider technology to be an approach to achieve more at work and improve productivity, these advantages are frequently affected with challenges such as the need for more time to execute projects, and pressure due to competition, as organizations want to accomplish more with less, taking an insight into Moore's law.

Artificial Intelligence will, without a doubt, cause numerous jobs to go into extinction. However, new jobs will emerge, and not all roles will be influenced in the same way. This book considers the in-depth impact of technology on every aspect of life (business, finance, human resources (HR), marketing, health, transportation, manufacturing, information technology (IT), innovations, etc.) and even the future. By and large, digital transformation can be a tremendous source of leverage and improvement in the business world. However, it might likewise increase pressure on personnel and affect their professional productivity, personal well-being, and career performance. Organizations need to understand both challenges and opportunities if they are to live up to expectations or deliver to their maximum capacity.

Work requirements will tend to become even more demanding. Workload per person will increase, driven somewhat by the need to react to quickly increasing amounts of continuous information from automation and artificial intelligence of various sorts and from more requests for innovative ideas and real-time implementation of these ideas. Globalization and the developing interconnectedness of industries and organizations will likewise increase the number of factors that determine the future impact of technology as it continues to transform. Time constraint is the greatest test. The capacity of technology to help people accomplish more with less, does not generally help executives save time.

Digital transformation is crucial at this point in evolution. Teamwork will now be the norm for greater achievements on a global scale. Work environments will require coordination between more people across multiple functions. Collaboration is the best way to make the most of individual expertise, respond swiftly to business problems and boost competitiveness. Businesses now optimize the use of specialists, freelancers and employees across functions which of whom must work together to achieve a greater purpose, further strengthening the gig economy.

Acquiring new technological skills is the ideal approach to propel proficient objectives. Notwithstanding the increase in artificial intelligence and automation, every single functioning capacity of a sector will understand the need to continue learning, improving and acquiring new skills through training and experimental models as they strive to accomplish their professional objectives. This is getting progressively imperative with the rise of millennials who are significantly more learned about technology now more than at any other time. So now, technologists will begin to focus on models that can augment human activities by innovation that can improve efficiency, productivity, customer service, and overall work fulfillment.

Technology tools will make it easier to respond to the changing market requirements while taking advantage of big data insights and their interpretations. They, additionally, can possibly enable the accomplishment of set objectives of the structure of new products and services, overseeing bigger teams, maximizing marketing potentials, driving specialty units and improving pay and execution. If used effectively, they can take advantage of big market disruptions. This is already happening. People who describe themselves as successful tend to make use of new technology tools to achieve their goals, anticipate their continued adoption and utilization, and to perceive themselves as having a positive effect on their career. They have higher desires for utilizing new technologies to accomplish more work in the meantime and to save more opportunity for vital and inventive work.

Visualize a world with a generation thriving in digital skills; where a refugee becomes a sought-after employee, a military veteran excels in his retirement career choices, and students can create sensors that can activate timely warnings for natural disasters. All communities have access to clean water. A game warden can use data to protect endangered rhinos. Drones can replant a rainforest. That world is here; that time is now. Welcome to the digital revolution.

Today, technology and data intelligence are allowing people to change the way we address and ultimately solve our most pressing social and environmental challenges. Digitization is leading to a greater understanding of the connection and interdependency between business, society, and the planet. Cisco's CEO, Chuck Robbins sums it up nicely: "What is good for the world and what is good for business are more closely connected than ever before." As a person embedded in that Cisco culture, I strive to inspire, connect, and invest in opportunities that accelerate global problem solving – empowering people everywhere to work toward eradicating poverty, unemployment, climate change, and hunger. Now, we have the opportunity to harness the power of the digital revolution.

However, implementing technological ideas is becoming harder. With expanded globalization and competition, executives also believe that the tasks ahead of us will become considerably more demanding and complex, involving a number of factors and still higher expectations from the customers. People can now do more in less time thanks to technology, and this makes working more flexible. Maximizing the potential of technology will require constant reskilling at every level. Experts will be required to have a better knowledge of their personalized expertise. With the use of temporary experts to achieve goals, talents will become a factor used when necessary in every function. Cultural changes are also essential. Organizations will need to introduce more flexible and collaborative working environments that appeal to millennials, whom they must attract and retain.

While the digital transformation provides tremendous opportunities, it is a massive undertaking. It will require an enormous change in the entire organization, Business Processes, People, & Technology. I have decided to write this book to share some experiences, as an effort to provide help and support to organizations and executives while they embark on the journey of digital transformation. As I meet many business and technology leaders, I learn more and more that the risk which comes with not moving fast, is way too high. There is no doubt in my mind that every single industry and organization will be impacted by technology one way or another. Some organizations will be transformed and disrupted way more than others, and some will be completely changed. The scary part is, some will entirely vanish, and be displaced, as many already have been over the past two decades. So, what is the way forward? How do we better understand technological trends to ease implementation of innovative ideas into real-life solutions? What does the future hold with respect to all determining factors of the global business sector? Join me as we journey into the world of what I call Tech

Evolution and Digital Transformation; as we discuss carefully selected properties of the global systems, while taking into consideration the relationships between multiple factors and variables based on logical and research-based evidence that creates a predictive model of global digital transformation conditions with recommendations and suggestions for solutions towards a better future. The impact on society in all of its parts will be paramount and the impact on the economy will be beyond any estimation or prediction. Despite the fact that several books have been written on this subject, this book takes a unique approach on how to start a meaningful digital transformation journey along with its challenges and opportunities. What better way to start a journey than from the beginning? In the next chapter, I will be taking you back in time to how it all started. For you to witness the various transformations and progress that have been made over the years in technology and the industrial revolutions.

# CHAPTER 1

## INDUSTRIAL AND TECHNOLOGICAL REVOLUTION

Humans are naturally curious, with the urge to always try to make life easier and better. This curiosity and necessity for a better life gave rise to the industrial revolution that began as far back as three hundred years ago. Although the industrial revolutions have been in stages throughout the past decades, there have been four unique revolutions to date. Currently, we are standing on the brink of a technological revolution that will fundamentally alter the way we live, work, and relate to one another. In its scale, scope, and complexity, the transformation will be unlike anything humankind has experienced before. We do not yet know just how it will unfold, but one thing is clear: the response to it must be integrated and comprehensive, involving all stakeholders of the global polity, from the public and private sectors to academia and civil society. How did we get to this point in the industrial revolution?

## THE 1ST INDUSTRIAL REVOLUTION

We have come a long way from the First Industrial Revolution, which dates back to the 17th or early 18th century. It was majorly felt in some parts of the world such as Europe and North America. It was a period when for the most part agrarian and rural social orders turned into more urban and industrial ones. The textile and iron industry, alongside the innovation of the water wheel and then the steam engine, were some of the main characteristics of the Industrial Revolution. This industrial revolution formed part of the history of ancient civilization since before then, an agrarian system was prevalent in Europe and America. Every home was practically involved in manual labor due to the fact that machines were not yet invented to simplify living conditions; anything individuals needed, they had to get done with their hands. The mode of transportation back then was through the use of animals such as horses or donkeys. There was a lot of cultivation done by people due to the fact that subsistence farming was in vogue. Life was quite stressful for the poor as they received little wages, and they experienced malnourishment and sickness. Households were saddled with the responsibility of making their own consumables; they made their own food, garments, and furniture. They had to work to get what they needed. As time passed humans sought ways to improve living conditions and reduce waste of resources. This was what gave birth to the industrial revolution, which marked the progress from an agrarian economy, which was based on producing crops and maintaining farmlands, into an industrial era, which shifted the popular vocation of farming into the use of machines. This gave rise to factories, industries and mass production of goods. In the late 1700s, the industrial revolution spread from Britain to other countries in Europe, America and other parts of the world.

The first industrial revolution in the mid eighteenth century is one with many consequences. Beyond reasonable doubt, it sets the pace for the present digital age. Starting from Europe — England, the world witnessed a radical takeover of economy by the classless agrarian population of England which eventually spread to other countries in Europe and the USA. Coming from an aristocratic and feudalist England, the exponents of this revolution are the merchants who moved their merchandise from one country to another over sea and land. These men acquired charters with nobles and were granted rights of incorporation. They consequently got political autonomy and once they could float independently, they slowly brought the conservative English population along into the revolutionary and capitalist game.

### Effects of the 1st Industrial Revolution

Socially, the first industrial revolution changed the disposition towards agriculture amidst the peasants. Traditional England was full of peasants whose disposition towards agriculture was that it was only meant for the sustenance of their nuclear families. There were little or no commercial agricultural activities involving the peasants, except in cases of scarcity where the aristocrats, who had the producing power, hiked the prices of food, which ultimately led to the starvation and death of the peasants. The first industrial revolution changed the orientation of farming—for—survival that was predominant at the time, and replaced it with a new orientation of farming—for—commerce. This new outlook served as a fulcrum for the beginning of the full revolution.

The effect of the peasants taking up agriculture was that it brought them on a leveled ground with the aristocrats in the economic scale. Not only did these peasants gradually storm the markets with their goods, they also began determining the prices of goods based on demand and supply functions.

Urbanization also came with the first industrial revolution. The coming into being of cities caused heavy migration. The peasants gradually started moving from their locations to more commercially comfortable places and as such came about the consequences of extra labor forces in the provision of other non-agricultural goods and services.

# THE 2ND INDUSTRIAL REVOLUTION

The Second Industrial Revolution was between 1870 and 1914, just before World War I. It was a time of improvement for previous industries and creation of new ones like the steel, oil and electric power industry. It also made use of electricity for large scale production. Major innovations during this era were the phone, phonograph, light bulb, internal combustion engines, and so on. Further revolution into the 18th century, new machines were designed to increase production thereby reducing human labor and errors. These machines were powered by coal, iron, steel, and steam. There was plenty of development and improvement in this era. About 150 years ago, the world witnessed the birth of internal combustion engine and electricity amongst others. Electricity made the world brighter. Building on Benjamin Franklin's work on electricity, famous inventor Thomas Edison invented the light bulb in 1879, which replaced the traditional lantern. For instance, the textile industry, before the first revolution which is also known as the industrial revolution, was made in people's homes with merchants fully sourcing all they required for their cottage industry from getting cotton from their farm to processing them into materials for garments and other purposes using their hands. But during the industrial revolution, they were able to use machines to help them weave, sew and dye which enabled them, with speed, to mass produce. Another example is the transportation industry with the rising production from factories and various industries. There was pressing need to improve the mode of transportation of goods and road networks, and this led to the discovery of railways and steamships. The rise in trade led to the development of communication and banking systems which began the technological revolution; the 18th century was truly an age of enlightenment. It transformed the world from a traditional agrarian society into the technological and information age.

## Effects of the 2ⁿᵈ Industrial Revolution

The factory system became predominant and totally replaced the simple tools of production for personal consumption. The second industrial revolution marked the total removal of the use of simple machinery in order to cater for the teeming populations across new cities. The factory system began in 1771 when Richard Arkwright built the first factory. This total change from simple production to massive production came in the second revolution.

The presence of factories ushered in a phase of exploitation for the workers in those factories. In America, this caused a lot of social chaos as these workers were forced to obey new regulations which disrupted social engagements.

Electricity and more technology also became available to all and sundry at a cheaper cost. The second industrial revolution also refined the taste of both the rich and the poor by offering them some new and easier methods of getting work done. This also led to the decrease of factory workers because less people were needed for production as time went on.

## THE 3RD INDUSTRIAL REVOLUTION

The Third Industrial Revolution, or the Digital Revolution, was the era of the progression of technology from analog electronic and mechanical gadgets to the computerized and digital innovation accessible today. The period began around the 1980s. Innovations during this revolution include the PC, the Internet, ICT, and many more. After the industrial revolution, there was a great need for machines which could do lots of calculations without error. This led to the invention of computers. The technology and internet revolution started about 30 years ago with digital communication becoming widespread in the latter half of the 19th century. This era gave rise to major inventions in science and technology including the introduction of the computer and the internet which gained massive adoption all around the world. The technology revolution pioneered the use of digital technologies while the 20th century took the world further into the internet revolution. The world experienced a shift from analog into digital technology, from the transmission of information with physical measurement into signal processing in binary form or numbers that could be stored in electronic devices. The transistor was invented in 1947 by a team of American physicists John Bardeen, Walter Brattain, and William Shockley. The transistor served as a substance that could control electronic devices. It pioneered the way for advanced digital computers, digital cameras, radios, phones, amplifiers, and other modern electronic devices.

After the success of digital technology, the personal computer gained mass adoption all over the world with leading sales from IBM, Microsoft, Apple II, and other companies that were able to sell personal computers across homes, schools, hospitals, banks, military, and governments across the globe. This increased the need for more functions on the computer such as the internet. Due to the rise in the ownership of the computer, there was need to connect users across the world. This led to the invention of the World Wide Web also known as the internet. In 1989 computer scientist Tim Berners-Lee developed a system that would enable the world to share information online; this was known as the World Wide Web. Although it gained mass adoption in 1991, the internet grew rapidly, and many businesses began to be listed. This gave further rise to web applications, digital marketing, internet browsers, email, web search, dark web, e-commerce, networking and social media platforms. What made the digital revolution standout among other revolutions was its concept of disruptive technology and innovation.

From 1967 onwards, the world has seen much more progress than it has seen for thousands of years. And the key factor in this geometric progress includes nuclear energy and internet. This has in so many ways changed the pattern of doing things, of interaction and also the direction of the future. The third industrial revolution birthed Microsoft, Facebook, Cryptocurrency, Ecommerce, Artificial Intelligence and other things. This means that the industries are being handed over to engines entirely and to those who are innovative. The telling effect, therefore, is that this present revolution demands more logical thinking, less physical interaction and more mental soundness.

## THE 4$^{TH}$ INDUSTRIAL REVOLUTION

Currently, we're on the cusp of the Fourth Industrial Revolution, or Industry 4.0. It's quite different from the three Industrial Revolutions that preceded it—steam and water power, electricity and assembly lines, and computerization—because it will challenge even our ideas on what it means to be human. The Fourth Industrial Revolution describes the exponential changes in the way we live, work and relate to one another due to the adoption of cyber-physical systems, the Internet of Things and the Internet of Systems. As we implement smart technologies in our factories and workplaces, connected machines will interact, visualize the entire production chain, and make decisions autonomously. This revolution is expected to impact all disciplines, industries, and economies. While in some ways it's an extension of the computerization of the 3rd Industrial Revolution (Digital Revolution), due to the velocity, scope and systems impact of the changes of the fourth revolution, it is being considered a distinct era. The Fourth Industrial Revolution is disrupting almost every industry in every country and creating massive change in a non-linear way at unprecedented speed.

These technologies are disrupting almost every industry in every country, and the breadth and depth of these changes herald the transformation of entire systems of production, management, and governance. Emerging technology breakthroughs have emerged in fields such as artificial intelligence, robotics, the Internet of Things, autonomous vehicles, decentralized consensus, 3D printing, quantum computing, nanotechnology, biotechnology, 5G technology and many more. The fourth wave of the industrial revolution is expected to see the heavy implementation of several emerging technologies with a high potential for disruptive effects.

Indeed, one of the greatest promises of the Fourth Industrial Revolution is to improve the quality of life for the world's population and to raise income levels. For those in First World countries, who already enjoy some of the benefits of a connected world as well as new products and services developed to take advantage of the technologies, we appreciate the efficiencies and conveniences provided such as booking a flight to getting movie recommendations. Our workplaces and organizations are becoming "smarter" and more efficient as machines and humans start to work together, and we use connected devices to enhance our supply chains and warehouses. The technologies of the Fourth Industrial Revolution might even help us to prepare in a better way for natural disasters and potentially also undo some of the damage wrought by previous industrial revolutions.

## THE SOCIAL AND ECONOMIC IMPACT OF THE INDUSTRIAL REVOLUTION

One of the characterizing and most enduring highlights of the Industrial Revolution was the rise of urban areas; over 80% of people lived in rural regions. Communities turned out to be enormous in urban communities. By

1850, without precedent for world history, more people have been living in urban areas than in the countryside. As different nations in Europe and North America industrialized, they excessively proceeded with this way of urbanization. By 1920, a greater part of Americans lived in urban areas. In England, this procedure of urbanization proceeded with persistence all through the nineteenth century. The city of London developed from a populace of two million during early 1840 to five million, forty years after the fact (Hobsbawm, Industry and Empire 159). This procedure of urbanization gave rise to industrialization with the focus on employing more laborers and keeping them closer to their place of work. Furthermore, the new modern urban communities progressed toward becoming wellsprings of riches for the country. Despite the development in the slums, in riches and in industry urbanization, there was still some negative impact on these neighborhoods. They were hopeless and contaminated with several outbreaks of diseases. Streets were sloppy and needed walkways. Houses were constructed close to one another, ruling out ventilation. Homes needed toilets and sewage frameworks, and subsequently, drinking water sources, like wells, were much of the time sullied with illness.

Cholera, tuberculosis, typhus, typhoid, and flu desolated through new modern towns, particularly in poor average worker neighborhoods. In 1849, this number reached 10,000. There was so much of death from diseases such as cholera in a quarter of a year in London alone ("Public Health Timeline"). Tuberculosis killed 60,000 to 70,000 lives regularly in the nineteenth century (Robinson). People who got medical treatment were administered by unprepared specialists and untrained quacks. Specialists still utilized cures prominent during the Middle Ages, for example, phlebotomy and siphoning. They prepared dangerous mixtures of mercury, iron and arsenic. They additionally supported substantial utilization of retching and purgatives, the two of which seriously dried outpatients and could add to an early demise, particularly among babies and kids whose bodies would lose water hazardously quick (Robinson). Despite the fact that there were more specialists in urban communities, the future was much lower there than in the rest of nation. Poor food, sickness, absence of sanitation, and destructive medicinal treatment in these urban zones devastatingly affected the future of people surviving the industrial revolution.

## IMPACT ON WOMEN AND CHILDREN

From numerous points of view, women endured more than men. In both the urban craftsman economy and the rustic rural world, women were generally viewed as assuming a similarly significant job as men. They were full partners in the family's mission for financial achievement. Their status changed because of the Industrial Revolution. Their work turned into an item to be misused. They were given the least gifted, most minimal paying employment. They were consistently tormented by both their supervisors and their spouses. From various perspectives, their work and duties multiplied. They were in charge of their occupations in industry, yet they were likewise expected to proceed with their customary jobs at home. They worked for ten hours in the industrial facility and proceeded for untold hours once they arrived home. It must be recalled that lawmen still controlled their families. Women had no political, social, or financial rights outside their homes. Child labor changed because of the Industrial Revolution. Children were required to help the family in the traditional economy, yet more often than not, they had been allotted errands that were not proportionate with their age. Much the same as their moms, children started to be misused by their supervisors. The most perilous task for kids in the factories was unjamming big material machines that wove fabric. Since their hands and arms were so little, they could venture into little spaces where the texture would in general jam. The foreman would not kill the machine but rather would demand the children reach in to unstick the jam, this resulted in extreme harm to children. All workers, male, female, and kids, were in the long run viewed as exchangeable parts. As the industrial revolution expanded, machines started to become progressively refined; business owners preferred the use of machines over manual labor.

## BENEFITS OF THE INDUSTRIAL REVOLUTION

The Industrial Revolution is a standout amongst the most critical events in mankind's history and profoundly affected numerous countries all throughout the world; its effects can even now be found in our lives today. For instance, the Industrial Revolution prompted a considerable lot of the accompanying: the development of communist developments and work developments, women's activist developments, urbanization, and our advanced purchaser society. The Industrial Revolution was a noteworthy defining moment in history which was set apart by a move on the planet from an agrarian and workmanship economy to one commanded by industry and machine manufacturing. The Industrial Revolution expanded the wealth and material abundance of the Western world, which led to economic and social change. Radical new schools of a financial and philosophical idea started to change the traditional thoughts that long existed, thereby giving way for all-round progress in the world.

## Industrialization 1.0

The First Industrial Revolution started in the eighteenth century using steam power and mechanization of the major production process. Production accomplished multiple times the volume than it did with the use of human labor. Steam power expanded profitability. The Industrial Revolution was followed by an agricultural revolution that greatly improved food supply while diminishing the measure of work required. Generally, the essential objective of agribusiness was to create enough food to end famine and starvation. Clover and turnips were two of the most broadly utilized methods in the agricultural revolution. Poor harvests occasionally led to the rise in the prices of food, which led to most workers being upset. Farmers had to devise various methods for harvests during various seasons. This radically changed the hundreds of years old model of the "open field framework", in which farmland was utilized in a semi-open style that kept farmers from developing strategies that varied from the conventional ones. As the population grew further, the issue of famine came back. This reality made workers question any proposed changes in the acknowledged strategy for cultivating. They were not daring individuals, so they commonly battled to keep up the present state of affairs. As time progressed, commercialization emerged, which gave rise to the market economy. With the introduction of the market economy, workers were seen as a factor in the production; they needed rewards and compensation in order for any business to gain the best advantage and significance in this enterprising economy. This made a desire for more prominent benefits for workers.

## Industrialization 2.0

The Second Industrial Revolution started in the nineteenth century through the revelation of power and mechanical production system generation. Henry Ford (1863–1947) took large scale manufacturing from a slaughterhouse in Chicago: The pigs dangled from conveyor belts, and each butcher performed just a piece of the task of butchering the pig. Henry Ford converted the standard production of automobiles and over time made more adjustments to the automobile business model. Before, one station collected a whole car. Presently, the vehicles were delivered in fractional steps on the conveyor belt; this was faster with very minimal cost. As in every single beneficial transformation, skills enormously decided the personal satisfaction. Craftsman had the most effortless time changing to the new financial world. The way that they had exceptionally excelled in the agricultural revolution era empowered them to adjust to the new machines in the factories a lot simpler than their farming partners. The modern economy had another arrangement of principles and time plans for normal workers. The workplace moved inside; however, the pace of the work changed definitely. Rather than driving a horse that pulled a wagon, the machines drove the workers. The periods of the year were never again applicable to the time spent at work. Men were currently expected to work twelve to fourteen hours every day, five-and-a-half days of the seven days in a week, throughout the entire year. This was hard progress to make. A large number of people, who had once been considered profoundly gainful as agrarian experts, were unfit to hold occupations in light of their powerlessness to conform to this new routine of working with machines. By the start of the twentieth century, electricity turned into an essential source of living. It was simpler to use than water and steam, and empowered organizations to think power sources to individual machines. In the end, machines were structured with their own capacity sources, making them increasingly versatile. This period, likewise, observed the advancement of various administration programs that made it conceivable to expand the proficiency and viability of assembling offices. Division of work, where every laborer completes a piece of the absolute occupation, expanded efficiency. Large scale manufacturing of merchandise and utilizing mechanical production systems was the other major contributing factor. Frederick Taylor acquainted methodologies of examining occupations that would advance workers and working environment techniques. Industrial Revolution itself was essentially determined by the rise of technology, which perpetually changed the substance of the world, driving us into the cutting-edge period. The outer burning steam motor-controlled railroads, production lines, and enlivened the inner ignition motor and the automotive business. Energy demands prompted power and electric-based appliances. Telegraph prompted the phone, the internet and mobile technology. There are various guides to propose the great strides humankind took in the field of innovation during and as an outcome of the Industrial Revolution.

## Industrialization 3.0

The Third Industrial Revolution started during the '70s in the twentieth century through partial automation utilizing memory-programmable controls and computers. Since these advancements, we are currently ready to automate the major production process without human help. Known instances of this are robots that perform programmed schedules without human intervention. Over the most recent of the twentieth century, the development and assembling of electronic gadgets, for example, the transistor and, later, incorporated circuit chips, made it conceivable to completely automate machines to replace operators. This period likewise brought forth the improvement of software hardware to benefit from the electronic hardware. Incorporated frameworks, for example, arranging material necessities, were replaced by big business asset tools that empowered people to plan, schedule, and track product flow in the factory. Demands to reduce expenses made organizations source for low production costs in less developing countries. The all-inclusive geographic scattering brought about the formalization of the idea of supply chain management on board.

## Industrialization 4.0

In the 21st century, Industry 4.0 interfaces the internet of things (IOT) with assembling procedures to empower frameworks to share data, break it down, and use it to direct intelligent activities. The advancement of new innovation has been an essential driver of the development to Industry 4.0. A portion of the projects originally created during the later phases of the twentieth century, for example, shop floor control and product life cycle management, were farsighted ideas that come up short on the innovation expected to make their total usage conceivable. Presently, Industry 4.0 can enable these projects to achieve their maximum capacity.

*Trials and Opportunities of Industry 4.0*

Just like every other revolution, the Fourth Industrial Revolution has a lot of possibilities of raising the income rate of the world and improving the personal satisfaction of products and services. Those that are benefiting from the benefits of this current technological revolution are basically those that can afford it and that have access to the digitized world. The technological advancements so far have brought about products and services that tend to provide total satisfaction to the end users. Now, people that can afford to request a taxi, book a flight, purchase an item, make payments, listen to wide collections of music, watch movies and play wireless games are remotely connecting to a large number of people, products and services with great speed from the comfort of a device with internet connection.

Technology innovations are leading to miracles in the supply, increasing productivity and efficiency over a long stretch of time. Cost of doing business will eventually drop as means of communication and transportation are getting more affordable, with increasing opportunities of entry into a broader global market space with enormous growth witnessed in the global economy.

Inequality might further increase which will cause a disruption in the existing labor force or create a new one altogether. With the rate at which workers are getting displaced by automated machines across all industries, this will eventually lead to a big difference in the productivity of sectors that still rely majorly on workforce over those that have the capital to invest in artificial intelligence. However, this might lead to the possibilities of people getting more professional skills that might lead to better and more rewarding job opportunities as the labor market will begin to require more of experts than artisans.

The drastic increase in inequity might be the greatest economic concern related to this current revolution. Those that will benefit the most are the physical, intellectual capital providers, innovators, investors, and stakeholders, which clarifies the increase in wealth difference between those subject with capacity to invest in trendy technology and those that invest in labor. The technological trends associated with this revolution are the primary reasons that wages have stagnated, or even diminished for a greater part of the populace in high-income nations. The interest for experts with high skills has increased exponentially while the interest for workers with less education and lower skills has diminished a lot. The outcome is the new dawn of a labor market with high demand for highly skilled and no skill-at-all workers, leaving those with average qualities out of the equation.

This clarifies why such a large number of the workforce occupying the middle class are feeling disappointed and frightful that their livelihoods and those of their children might remain to stagnate or no longer exist in the nearest future. This revolution is bringing about an economy where the influential and highly skilled take all while offering just restricted access to the middle class due to the philosophies of democracy and the right to natural resources.

The universality of digital technologies and the uniqueness of social media further fuel the general trends in this revolution. With about thirty percent or more of the world population having access to social media platforms, there is an increase in the way people learn, share information and connect to others all over the world. A person that has not been to a part of the world before might be able to school another person that has spent all their lives in that part of the world. The access to information is now limitless. In an ideal situation, the unlimited access to relate with others leads to an increase in cross-cultural cohesion and comprehension. However, this can also lead to an increase in unrealistic expectations based on the standard of living in other parts of the world, with respect to what people see as success for a person or a group, with the possibilities of propagating extreme ideologies that can significantly influence the way of life for people.

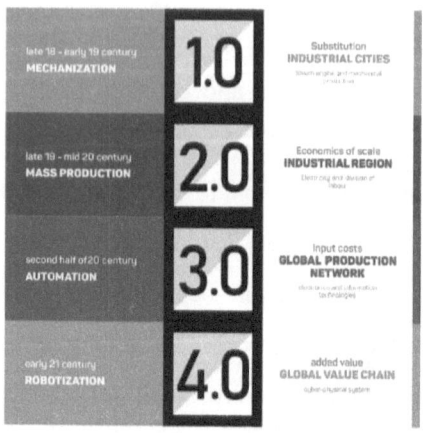

*Impact of Industry 4.0 On Business and Economy*

The main topic of discussion around global CEOs and senior business administrators is that the velocity of innovations and the speed of interruption are becoming difficult to predict or understand, and that these results are turning into steady innovative surprises, where those that are believed to be well informed and knowledgeable are caught unawares. It is clearly visible that the technologies evolving in this Fourth Industrial Revolution are cut across all industries and with an impact on virtually every business.

With respect to supply, numerous ventures are seeing the advancement of new technologies that introduce new approaches to meeting already existing necessities and fundamentally disrupting the existing industry value chains. We are likewise going to experience disruption in the way agile and innovative competitors emerge, due to the access to worldwide research, improvement, promotion, deals, and supplies. Already established stakeholders that do not follow the trends can easily run out of business when there are improved quality, speed, and cost from new arrivals. We are likewise experiencing changes in the customer behavioral patterns, with unique and ever-changing needs on demand, driving changes in the way organizations configure, advertise and deliver products and services. There might be increased social tensions as a result of the socioeconomic changes brought by the Fourth Industrial Revolution.

All in all, there are four primary impacts that the Fourth Industrial Revolution has on business. The impact on expectations of the consumers, on the enhancement of products and services, on collaborations for innovation purposes, and on organizational structures. Consumers are progressively at the focal point of the economy, which is tied in with improving customer's satisfaction. Products and services are now improved with technology capacities that increase their value. Technology is helping to improve the durability and reuse-ability of products, while feedback and analysis are changing maintenance procedures. All these and more are achievable with Industry 4.0, but given the speed at which these innovations and disruptions are occurring, we seem to be lacking behind expectations. However, the rise of global platforms and new ways of doing business, at long last, implies that skill, culture, and organizational structures should be re-evaluated.

By and large, the rigid transformation from basic digitization of the Third Industrial Revolution to this advanced mixture of innovative technology of the Fourth Industrial Revolution is compelling organizations to reconsider the way they function. The overall concern, in any case, is that business pioneers and executives need to comprehend their evolving working conditions, challenge the suppositions of the way they function, and continue to bring forth innovation, no matter the level of disruption.

With the rise in the rate at which citizens are able to stand up against governments, emerging platforms through technology are progressively empowering people to engage with their governments, voice their opinions, organize their endeavors, and even over-rule the policies of demanding authorities. At the same time, governments, with the use of technology, are increasing their authority over populaces, due to the use of reconnaissance systems and the capacity to control institutionalized digital frameworks. However, government structures will be forced to progressively change their present ways of policymaking and public interference, due to the possible decentralization of power that this industrial revolution brings to the table.

Eventually, the capacity of government to adapt to the trends will determine if they are able to cope with handling public authority. If they are able to comprehend the new universal disruptive troublesome changes, having better, effective and transparent policies that will empower them to keep holding on to power, they will persevere. A government that is unable to move with the changing tides will face a lot of oppositions.

This current revolution will be more visible in the way it affects policy-making decisions. Existing frameworks of governance evolved with the Second Industrial Revolution when those in authority had sufficient resources to look into particular issues and fundamental reaction to structure current policies governing most of the world. The entire procedure was intended to be direct and strictly following a particular road map. In any case, such approaches are no longer usable. Given the Fourth Industrial Revolution's fast pace of progress and wide effects, governments around the world are being faced with a lot of challenges and with all indication, so many are unable to adapt to the disruptive nature of the innovation, which is obvious in the amount of increase in the level of protest and demand for certain policies by citizens around the world.

The solution to this can be as simple as the government of nations around the world beginning to become more open to the innovations that come with this revolution, implementing policies that accommodate the rate at which changes are happening and preserving the interest of the common man. How, at that point, would they be able to protect the interest of people while making efforts to accommodate technological and industrial innovations? By implementing 'Agile' policies, similar to how the private sectors have progressively adopted 'Agile' response to improve customer satisfaction. This implies that those in charge need to persistently adjust to new and quick evolving conditions, preparing themselves to comprehend what it is they are facing genuinely. To do such, government agencies will have to work together intimately with business and society at large.

There is a ripple effect likewise on national and international issues on security, disrupting the likelihood and the way conflicts are happening. According to the history of security and global warfare, technological revolution plays a huge part and the current revolution is no exemption. Current clashes including countries are progressively becoming a hybrid with the combination of combat and intelligence systems. The difference between the intent of peace and war, terrorism and militancy, with respect to cyber-warfare is becoming difficult to say for certain.

The tendencies for the simplicity of the use and production of weapons of mass destruction caused by this current revolution have given rise to even more significant technological breakthroughs in the security sectors to prevent and control the use of this weapon both physically and via the cyber for the act of terrorism. This means for every bad that the revolution brings there is a greater good to counter the adverse effects caused by the drastic rate of changes happening all over the world.

*The Impact of Industry 4.0 On People*

The Fourth Industrial Revolution is disrupting not only our business and environment but also our identity. It influences our personality and defines who we really are: our privacy, the way we work, our sense of customer satisfaction, our relationship with others, our careers, our personal and social livese and so on with tremendous changes in the health sector, and likewise disrupting human development patterns. The level of impact it has on us is limitless due to the levels at which our imaginative ideas can be brought to life.

We no longer engage in personal and meaningful discussions with loved ones as the smartphone has taken over that aspect of life of virtually everyone that has access to a phone. It can be incredible what technology has brought into our lives, but what about taking time to pause and reflect on our current life experiences. Are we truly living up to our full potentials as humans when it comes to our relationship with ourselves and others?

The issue of privacy is one critical component of this technological revolution. We are aware of how important privacy is but are yet to help improve performances of products and services; our privacy is intruded for the gathering of information and tracking purposes. The demand for the controlled intrusion of privacy is already on the rise as people are no longer comfortable with the ability of most AI based platforms knowing virtually everything about their privacy. Also, with Artificial intelligence and biotechnology, people are beginning to to alter their physical and mental features, which will go a long way to redefine what it will mean to be a human being in the nearest future.

*Possibilities in The Future*

The level of disruption by this revolution is still very much under our control as we are the inventors of this innovation. Yes, we! We are all responsible in one way or another for the exponential changes happening all around us. With the way we shop, work and make investments, everything tends to add up to what is causing this evolution. Change is inevitable, it is the only constant thing on earth, so we have to be able to have control over it. We should be able to build on this Fourth Industrial Revolution and develop a future that reflects every one of our regular goals and aspirations as humans.

For us to achieve this, we must be able to have a collective perspective of how we want technology to affect our lives and the way it affects all other fundamental elements associated with us. We have never had a better revolution in the history of the industrial revolution. But those in the position of authority are making decisions and policies based on an already extinct revolution. We need innovative thinkers that can make use of agile techniques to respond to the disruption that comes with this revolution.

The future depends on what we shape it to be. Even as Artificial Intelligence and other innovations are aiming at robotizing humanity, we are the ones still in control of the circumstances of the existence of these innovations.

Technology should not be the one to define us, we should be the ones to define technology.

All stakeholders need to effectively plan for and regulate this innovation to ensure the safety of the world at large. Typically, first-adopters of technology are the ones with the financial means to secure it, and that technology can catapult their continued success increasing the economic growth. Some jobs will obviously go extinct. Then again, the changes might come so fast, that even those who are already at the forefront, in terms of their preparation and knowledge, might not be able to keep up with the ripple effects of this revolution. Even at that, the World Economic Forum in early 2019 noted that Society 5.0 might be the next wave after the Fourth Industrial Revolution (Industry 4.0), and we are yet to fully comprehend the current revolution. The next chapter dives deep into the world of digital transformation, which is the remarkable center piece of Industry 4.0. What is digital transformation all about? The answer to this question will clear your doubts and misapprehensions about our current revolution. Find out more about this in chapter 2 below.

# CHAPTER 2

## DIGITAL TRANSFORMATION

As clearly established, digital transformation is the center piece of the current industrial revolution. That is our reality today. Digital transformation can be said to be the use of existing technologies, ideas and innovations to radically transform the scope, strategies, structures and general overview and productivity of a business to meet the demands of the present as well as prepare for the future. This means quite different things to each industry; however, the goal of digital transformation is the same across industries.

For a farmer, digital transformation may mean the use of IoT sensors, robots, drones, trackers on machines and the likes in order to improve the security of their businesses and to maintain a high level of productivity in order to meet the present need as well as prepare well against the need to increase the world productivity of agricultural products by 60% by 2030. It is not surprising that the industry has spent over $4.8 billion on digital technology investments since between 2015 and 2019 alone.

For the fashion industry, digital transformation has to do with customer centered services, digital advertising, augmented reality, market-based production, brand building, 3D prototyping and designing, right modeling and customer care services — all on the digital space. The digital transformation in the fashion industry might just be another tremendous move in the digital world that one should not fail to commend. The present demand for fashion wears are mostly known through digital surveys and other enhancing methods.

The promises of digital transformation are a better today and an assured tomorrow. Every digital transformation agent is built for speed and agility, built to manage complexity, and designed for large scale traffic. And this cuts across various areas of life.

Digital transformation albeit problem-solving, also creates problems. The greatest of this is that digital transformation makes the machineries become the nucleus of the universe in the place of humanity. That explains why we would have robot doctors instead of humans who have fulfilled the academic qualifications. Digital transformation would thereafter render some industries useless and some people permanently unemployed. So, in essence, it is going to create massive unemployment and lack of occupational organizations.

Digital transformation has rapidly brought significant changes in human procedures, skills, and structures by transforming societies and empowering individuals with seamless technologies in every aspect of life. Digital transformation has helped businesses, organizations, public sectors and governments tackle tough societal and world issues with the use of emerging technology issues like pollution, aging population, loneliness, unemployment, diseases, food scarcity, earthquakes and several others. In a country like Japan, which is highly advanced in technology, the government has rolled out the industry 5.0 which makes the country one of the greatest technological hubs in the world. Areas of digital transformation include the advancement of skills, productivity, creativity, innovation, customer needs, industries, and communication that has enabled people to tap into new data, opportunities and generally advanced living standards. Operational adaptability and advancement are key drivers and objectives of technological change. The human factor is key in all dimensions of digital transformation in the phases of change. The human factor is crucial in all the stages of digital transformation. Some individuals do not fully embrace digital technologies because they value human interaction compared to interfacing with technological gadgets. The goal of digital transformation is the introduction of speed, accuracy, innovation, and reduction of stress and errors by humans. Innovation drives the requirement for advanced change and supports the digitization of global trends.

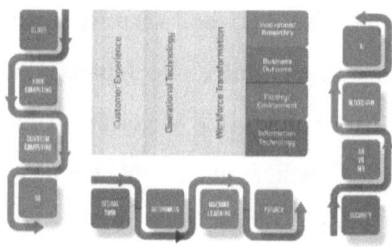

## WHAT IS DIGITAL TRANSFORMATION?

With the mass adoption of new digital technologies such as social networks, mobile, big data, and several others, firms in practically all industries are directing different activities to explore and enjoy their benefits. This, much of the time, includes changes in key business operations which influences products and procedures, just as organizational structures in organizations need to build up the board practices to oversee these complex changes. Consequently, the general public is facing quick and radical change because of the development of digital technologies and their omnipresent infiltration in local and global markets. Organizations are facing harder challenges as a result of globalization, which puts a strain on them to go digital before others do. They have to be alert in order to survive and gain a competitive advantage before their competitors do. Digital pioneers such as Amazon, Facebook, and Google have grown into incredible companies, while organizations that were market leaders in the traditional business model have considered digital transformation a major threat and that has made them lose their customers to these digital giants. In spite of the varieties of technological innovation and plans for their usage, regardless of whether in business, public, administration or private life, genuine Digital Transformation is taking much longer and confronting more difficulties than has been anticipated. For the success of digital transformation, organizations need to build up a wide-scope of abilities, which will change their business operations and give them constant competitive advantage. Digital technology needs to become fundamental on how the business works, and the possible need to reevaluate and re-design plans of action in order to stay aggressive. Wikipedia defines digital transformation as changes related to the use of digital technology in all parts of human society. This definition goes past business and incorporates all effects on society, regardless of whether it's training, government, sports, expressions, relaxation, or so forth. According to Altimeter, digital transformation is the realignment of, or new investment in, technology and business models to more effectively engage digital customers at every touch point in the customer experience lifecycle. Capgemini and MIT Center for Digital Business defines digital transformation as the use of technology to improve performance or reach of enterprises radically. While their definition may look oversimplified, the power is in the information behind the definition. It's a worldwide investigation of how 157 executives in 50 large traditional organizations are taking advantage and profiting from digital transformation.

## MIT Sloan's 9 Elements of Digital Transformation

In The Nine Elements of Digital Transformation, George Westerman, Didier Bonnet and Andrew McAfee recognize the key traits of digital transformation change. The nine components are excerpted from their advanced report: Digital Transformation: A Roadmap for Billion-Dollar Organizations.

**Customer Understanding**: Customer Understanding is where organizations are beginning to exploit as opposed to past interest in investment decisions. Organizations now prefer to have a deep insight into the demography of their customers in order to develop the best products for each market segment.

**Personal Selling**: Top-Line Growth is the place organizations are utilizing innovation to improve face to face sales discussions.

**Customer Touch Points**: Customer Touch Points are helping improve customer service by upgrading to digital activities to help organizations always stay in touch with their customers online or offline.

**Process Digitization**: Digitization can empower organizations to refocus their interest in more strategic tasks that would benefit the organization greatly on short, medium, and long-term periods.

**Employee**: Individual level work has generally been virtualized thereby, isolating the work procedure to be done by employees from the location of the work.

**Performance Management:** Company executives now have deeper insights into products, regions and customers, thereby allowing choices to be made on genuine information and not on suppositions.

**Modified Businesses**: Digitization has helped business owners expand their businesses physically and digitally by allowing them to share content across various platforms.

**New Digital Businesses**: Organizations are using various online platforms to introduce digital products that go in line with their traditional or physical products.

**Digital Globalization:** Organizations are gradually changing from local operations into global and multinational operations, thereby selling their products worldwide.

## Digital Transformation in Organizations

Digital transformation has been happening in organizations since the 1950s. The vacuum tube PC (personal computer) of 1943–1958 prompted changes in bookkeeping, which has changed the process of accounting today. The additions were humble due to technology constraints and limits. In 1960 there was more transformation included in the manufacturing of robots, online transactions and the time used in carrying out various transactions. By the mid-1970s, the PC transformation was just starting. All through the 1980s, the mass adoption of computing technology quickened. Visicalc and Lotus123 were the best applications that changed administration and the executive's basic leadership. The 1990s brought about data warehouses, local area networks, the worldwide Internet, digital data storage, and digital phones to the growing innovation conceivable outcomes accessible to managers and supervisors. The 2000s saw the realization of affordable mobile phones, quicker parallel processors, distributed computing and storage and digital cellular networks. Digital data storage and computing capacities expanded exponentially in the mid-2010s. Enterprise applications, the Internet of Things (IoT), Machine Learning, Artificial Intelligence (AI) applications, speech recognition, and analytics technologies were able to provide real-time monitoring and assistant and personalized support to help organizations in making the right decisions based on predictive analytics.

## Literature Review on Digital Transformation

From an advanced period, point of view, DT stresses the key change in our reality due to the inescapable nature and multiplication of advanced innovations (Anderson and Lanzolla 2010). Seemingly, we have achieved the fourth industrial revolution, which expands on new digital technologies with full power, whereby both the improvement and dispersion of advancements are a lot quicker than previous industrial revolutions (Schwab 2016). A new worldwide world economy, described by dynamism, customization, and exceptional rivalry, is creating the foundations for prevailing in the digital transformation era (Atkinson 2005). Moreover, the novel idea of a round or sharing economy is moving assets to a model where materials, energy, work, and new data connect and advance human life (Schwab 2016). The disruptive nature of digital technologies has revolutionized the way that industries operate and how the traditional boundaries between them have been dissolved. In recent years, manufacturing has gained popularity with the introduced concepts of "Industry 4.0", "smart factories" and "advanced manufacturing", which seek to enable industry to navigate its way through digitalization through the use of cyber-physical systems in the production network and service-orientation in traditional industries (Lasi et al. 2014; Blau and Gobble 2014). New technologies have also accentuated the changing network dynamics from the center of organizations to accommodate digitally-engaged customers at the edge, where consumers and communities co-create value in a digital ecosystem (Gray et al. 2013). Value network competition is another research area for academics who seek to explore how IT affects overlapping as well as non-overlapping, networks (Evangelos Katsamakas 2014). The need for transformation is also a clear business reality, which occurs in all industries and impacts companies of all sizes and shapes (Basole 2016). It is no surprise that "90% of business leaders in the U.S. and U.K. are expecting IT and digital technologies to make an increasing strategic contribution to their overall business in the coming decade" (Hess et al. 2016). DT is also exhibited in the "extended self", where technological changes dramatically affect the way in which individuals present themselves and communicate. Five crucial changes resulting from the digital age have been conceptualized in the literature (Belk 2013): de-materialization of possessions in the form of photos and videos, re-embodiment of our physical bodies into pictures and videos, sharing more with the help of digital devices, co-constructing the sense of self through digital enablers such as social media and blogs and a distributed memory, where human memories are outsourced to engines and hard drives. This individual level of digital transformation allows an exponential increase in digital data volume, revealing huge amounts of information floods that often bypass intentionally constructed barriers. Thus, researchers have recommended ways of managing the super-transparency of people in our world today (Austin & Upton 2016).

### Digital Transformation in Marketing.

At an abnormal state, the objective of advanced change in digital transformation has, over the years, helped organizations increase their customer base at a minimal cost. Companies have discovered more customers through digital marketing while spending less on advertising. Digital marketing has been said to generate more leads for companies and have helped organizations draw nearer to the majority of their customers. The shift from analog to digital advertising materials is commonly less expensive to create and circulate. Email, specifically, is far more affordable than the traditional forms of advertising. Rather than arranging a one-estimate fits-all down the marketing funnel, advertisers can observe 1-to-1 journeys that observe customer behavior and shape the experience to best suit every individual purchaser. Digital transformation helps advertisers to interact with individual customers, take a look at shoppers as people and examine their conduct from the first touchpoint entirely through the purchasing process. Salesmen especially, gain advantage from the access to more and better information about customers. Online networking is all over, crushing up news, excitement, and brand communications. PricewaterhouseCoopers, as of late discovered that 78% of shoppers were here and there affected by internet-based life during their purchasing procedure. What's more, almost 50% of shoppers said their purchasing conduct was straightforwardly influenced by surveys and remarks they ran over on social media platforms. The Digital Marketing Institute relevantly stated, "Effective social dealers can be viewed as idea pioneers, or even confided in as advisors, by customers as they offer some benefit through industry experiences, and offering answers for regular buyer inquiries".

## Digital Business Transformation

Digital innovations have changed our lives, work societies and businesses. The accelerating pace of change in business transformation has brought about rapid disruption in regions and business ecosystems. Technological innovation has one way or another transformed business transactions, business meetings, trades, business communication, management, vision, mission, company culture, reporting, and general business administration. For instance, the traditional business model of opening numerous company branches in several places to serve a larger customer base has today been streamlined to having a multinational corporation with the various franchise in different regions based on the influence of digital transformation. Innovative technologies ranging from the cloud, big data, artificial intelligence, analytics, mobility, the internet of things and emerging technologies serve as enablers in digital transformation. However, customer behaviour, needs, and perception have shaped the emergence of the role of emerging technologies in business operations.

**Areas of digital Business Transformation**

Digital business transformation cut across all areas in business operations such as human resource, marketing, security, finance, customer service, and all business administrations. For instance, in the area of marketing the old school technique before the advent of computers was face to face salesmanship. Billboard advertising, and other techniques were relevant many decades back, but today it is a different ball game as companies use digital marketing techniques to reach a diverse customer base across the globe.

Digital transformation has greatly influenced the business process in areas of optimization, process management, business automation, business goals and design. In today's business world, robotics process automation is an emerging technology in which the core aim of organizations is having software robots and artificial intelligence workers. Every business wants to reduce costs and maximize productivity thereby giving rise to digital transformations in business sectors.

Organizations' business plans have drastically changed over the past few decades. From the go-to-advertise approach and incentive to the manners in which it looks at profit and successfully changes its business models, taking advantage of novel income sources and strategies, digital transformation has made organizations embrace change by dropping certain traditional business practices such as bulky paper documentation into better digital ways of storing data and information.

Business networks have transformed over the years with technological innovations. Business ecosystems have become more connected with customers, partners, stakeholders, governments and other players in the internal and external environment of businesses such as regulatory and economic bodies who one way or another work to create a smart city. Digitization has enabled businesses to get the right data and information to help them develop the best product that is needed in regions and communities in the business ecosystem.

Business executives do not just rely on traditional assets like land, buildings, and machinery, they now embrace digital assets like email accounts, databases, virtual properties, virtual currencies and data as the backbone of the business. Digitization has made it possible for business executives to view clients and data as genuine resources in all points of view. Thereby, they try to secure major information in this regard.

Digital transformation has greatly influenced company culture as businesses now integrate coordinated efforts in business endeavours. Businesses today make collective business decisions, embrace change, and use different strategies which has prompted new plans of actions and profitability for organizations.

## Digital change in the Retail Business

Retail is a standout amongst the most quickly changing verticals over the world and is frequently at the bleeding edge of technological headway to keep pace with the advancing needs of diverse customer needs. Retail business has experienced digital change from information and data improvement, payment systems, inventory, return policy, and shipment of goods where client desires are requiring changes and upgrades with respect to consistent improved customer experience in every area from online purchases to product delivery.

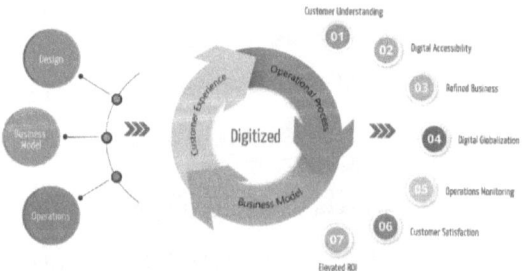

## The Economy and Speed of Change

Digital transformation has, over the years, influenced growth in the economies of different countries. Technological advancements have made it possible for governments to have the right data on their countries population, gross domestic product, gross national product and other economical concepts which have enabled them to enact policies that are best fit in various economic sectors; this exponential development and pace of progress is only a small amount of what is yet to come. The reception and chances of innovations under the umbrellas of social business, cloud, portability, big data, psychology, the Internet of Things and more, will dependably accelerate changes crosswise over society. Exponential development or speed of progress in any region can occur at the most unforeseen minutes, making the conditions to be prepared for, fast advancements and in a perfect world.

## Digital Transformation in Government

The job and structure of national and nearby governments, government offices, state-supported associations, and public establishments vary from nation to nation, where governments are included. For example, open social insurance, transport, open foundation, policing and safeguard, resident or administrations share a number of traits in the difficulties and needs not necessarily only from the advanced change point of view. A digital transformation like e-government privacy laws, in numerous different territories, are key in the digitization of procedures and the undertaking for the government. The main driver of advanced change in government and the open area is cost investment funds. In reality, as we know it, where populaces are maturing, a blend of national and geopolitical movements require decisions and changes. Digital Transformation is front and centre to the strategies of governments around the world. Most leaders realize the mega forces of disruption that are fuelled by technologies; they see Digitization as a game changer that would accelerate countries' priorities, such as increasing GDP, creating jobs, enhancing the quality of lives, safety and security, healthcare & education, and so on. Some countries are leading the trends, for example two years ago, UAE appointed a ministry for Artificial intelligence. In Saudi Arabia, the ministry of communications and information technology MCIT, and the National Digitization Unit (NDU), are leading and building the foundation for digital transformations that will help in realizing the country's 2030 vision; the e-government services are considered to be some of the most advanced in the world, they provide massive simplicity, and provide highest citizens' satisfaction. It is the case with most countries that believe digitization is important. They have created a government organization to lead digitization for the country, i.e. Denmark has the Danish Agency for Digitization and in Estonia, 99% of services are offered online.

## Digital Transformation in The Health Sector

Decades back the plan for a medical check-up had a lot of waiting time; from consultation to getting results to getting prescriptions from the pharmacy, the process could take several hours. Then, it would take days to get results back. Driven by the requirement for a superior client experience, health-related services are encountering an enormous move in advanced medicinal services as innovation is helping us live more with increased security and more beneficial, progressively profitable lives. Digital transformation in health is opening the genuine capability of human DNA examination by empowering really customized testing and treatment that could immensely improve quiet results for an enormous scope of sicknesses. Wearables that are Savvy with screens that can gather customized, ongoing information are empowering more beneficial ways of life, and gathering reams of information to sustain into medicinal research. A few organizations have just acquainted wearables into the working environment with lift execution. By observing the feelings of anxiety and wellbeing of their staff, they promote higher efficiency.

## Digital Transformation in Transportation

Digitization is affecting radical changes in the transportation and logistics division. Utilizing new innovation will change business and operational procedures. Effectiveness, advancement, speed, and timing have dependably been pivotal in logistics and transportation. With the rise of smart products, customers are requesting a more prominent dimension of administration from their transportation suppliers, the business is perceiving the need to change or else wind up outdated. Today, in the midst of the scope of quickening advancements where advanced changes are influencing the following disruption of the industry, known as Industry 4.0, it is considered important for transportation companies to stay in trend with the latest technological innovations in their sector. Setting up an advanced front end that could possibly furnish clients with an advantageous one-stop shop involvement, would improve operational efficiency. Cutting edge Solutions, looking at future operational upgrades by means of transportation technology, man-made reasoning, and even enlarged reality can help further raise strategic operational efficiency in dissemination, warehousing, picking and delivery in the transportation sector.

## Digital Transformation in the Financial Sector

The financial industry is evolving; decades back banks were more interested in having a physical branch where customers would go into a branch to get the majority of their financial needs. Automated teller machines (ATMs), as a second channel, became popular in the 1980s. The web made different channels available for customers to draw in with their banks with various ways to meet their financial needs using their gadgets. There are a lot more gadgets developing, which are holding customer information; this information is being transformed into a knowledge database that helps make more improved financial decisions. In the present advanced financial era, a lot of financial transactions can be done with a smartphone from money transfer to loans to payments of utility bills and many more. Financial services are now becoming seamless. The days when digital currencies kept up an unsafe presence on the edge of the money related world with many people showing great fear and concern about them are long past. Today they are seen as authentic, very advantageous, and valuable tools that anyone can use for a variety of purposes, from doing regular financial exchanges to making speculations.

## Digital Transformation in Education

Technology empowers the student to learn from a global world perspective, integrating several learning approaches that can help them get the best learning outcomes. For quite a long time, numerous instructors seemed to feel that learning and playing were fundamentally unrelated. Be that as it may, amusement components can tackle issues teachers have looked at for ages. They give quick input for activities and enable students to learn, for instance, gratification empowers students to utilize their insight by tackling issues in environments copying reality, and this could incredibly build inspiration and commitment in the minds of students as the instructive segment turns out to be progressively focused, turning into a vital method for survival as this new advanced world expects instructors to adjust and embrace technological innovations, approaches, and outlooks. On a progressively large-scale level, tertiary instruction enrolment rates all around are required to rise quickly by 14 million new students each year from this point until 2030. This would require 13 new institutions to be assembled each week, meaning 700 every year, each serving 20,000 students in the event that they are for the most part going to be instructed face-to-face. There is a need to enhance in scaling the supply of instruction so that individuals, edgy to learn, are not denied the opportunity to follow their dreams.

Each student is special, with their very own qualities and shortcomings; numerous instructors have commonly had minimal decisions on the teaching pattern that is appropriate for each student. Most instructive procedures have remained for all intents and purposes unaltered for quite a long time. The advanced disruption has not overlooked the educational sector, schools or universities which have at long last begun presenting extreme changes, and throughout the following decade, we are probably going to see a digital transformation. Students currently approach huge stores of learning material and courses to choose from a considerable lot of free online colleges around the globe with the flexibility to study wherever and at whatever point they need. There is access for experts and more seasoned students, and individuals from nations everywhere throughout the world. Learning on the web is fun, agreeable, intuitive, vivid, associated with the web, and information is driven. Capabilities are worked around the student. Web-based learning is additionally socially intelligent, enabling students to interface with each other everywhere throughout the world.

After getting to fully understand what Digital Transformation is all about and how it is having remarkable impacts on sectors across the globe. The next phase is for us to understand those factors contributing to the digital transformation responsible for the magnitude of technological disruption that we experience in the world today. How do we get to understand the magnitude of disruption caused by this revolution? We will be discussing the forces responsible for these disruptions in the next chapter titled Mega Forces of Disruption.

# CHAPTER 3

## MEGA FORCES OF DISRUPTION

The world is gradually coming to the age of disruption in almost every sector, from agriculture, finance, technology, education, manufacturing, tourism, medicine, and others. The intellectual capacity of humans over the past decades has grown; more so why technology has greatly advanced to help improve the standard of living. For example, automation and driverless vehicles are changing supply chains and coordination. Business, economic and political policies are also being disrupted very rapidly. Most of the forces that gave rise to disruption emanated from the essential powers of innovation, globalization and demographics. They developed in waves and each new wave gave rise to high need for improvement (innovation), growing influence (globalization) and aging (demographics), which has advanced the cause of megatrends in various sectors, work, consumer perception, food, regions, culture, geography, health, communities, design and general human behavior that has continually driven the need for a higher standard of living. Technology and innovation have constantly enlarged human abilities by helping people perform tasks seamlessly. Computerized reasoning, such as artificial intelligence (AI), augmented reality (AR), virtual reality (VR), biotechnology, nanotechnology, 3D printing, blockchain and other technological trends have brought forth major transformation and advancement in our human lives by enhancing interaction, visualization, connection and several endless possibilities in both physical and digital worlds.

Prior to the establishment of Intel Corporation in 1968, Gordon Moore made his assertion, which is referred to as Moore's law, which states that every two years the power of computers will double which would give rise to supercomputers with accelerated speed in technological improvements. Computational power and efficiency would keep advancing every single day. Today, we have many super-fast computers that are available to a wide customer base at an affordable price. For instance, a 3.2 gigahertz processor conveys about 3.2 billion pulses each second. This is the secret behind computing power. It all comes down to how fast a machine can perform. Moore's law gave us an insight into digital disruption. Today we do not just have supercomputers but smartphones which are smaller and perform very quick tasks as well.

Technology gave rise to major disruption in areas like social media, which was a major event changer from the traditional way of communication, smartphones, cloud technologies, internet speed, data and analytics. The advancement of the internet and smart devices gave rise to connecting several homes or office devices such as toasters, coffee makers, fridges and several others through the Internet of Things or IoT to be able to interact in the digital age.

Before the internet came, some persons were regarded as indispensable based on the enormous value of the services they rendered. Looking at the present reality, we could see that some of these people and jobs are no more in vogue, while others have benefited immensely from the largeness of the internet. So, picking just a few of the many, I'd like to pitch pre-internet industrial services vs internet-based services.

Mailman vs Email — Those who have lived before the internet came through and are still around could tell us how important the mail service is. This way, you wake up every Saturday morning with a key in your hand and walk across the field to your box, opening to see if you've got a mail from a friend far away or a son in college. Then when you write a mail, you walk or rather board a bus to the mailman's office where you trust him with the duty of delivering your message. However, since the arrival of the internet, sending a mail could just have become the easiest thing to do. Instead of the hectic routine of sending and receiving mails, you just get a cellphone or a computer and punch certain keys on your Gmail or Yahoo mail and send your mails with a cup of coffee on your bedside table. Internet made mailing easier, but it also disrupted an entire industry. The post office is of little or no relevance today, and that has already incapacitated some of whose lives have been lived with that synchronization. All thanks to the internet.

Newspapers & Media houses vs Cellphones — to belittle the importance of the newspaper, radios and television channels would have been a mad man's speech a few decades ago, but today, those industries have been forced to evolve or die by the presence of cellphones and computers who can access news directly on websites and also make their adverts on the e-world as well. The media houses and the likes are an endangered species with the emergence of the internet.

Uber vs Street taxi — Years ago, one had to wait for the next taxi, probably at a crowded bus stop in Hong Kong or other highly populated cities. But with the help of the internet, online companies like Uber came through and changed the story of the transport system forever. Those with disabilities can stay in their rooms and request for a taxi. This makes transportation easier to a reasonable extent.

Amazon vs Bookstores - Internet made it possible to purchase books a million miles away. This is to the detriment of those who operate bookstores — but not only that, a shopper would also buy goods on Amazon and ship it at no additional cost. This makes it not only a revolutionary departure from the traditional retail of books, but a major disrupter to the entire retail industry.

We could also pitch crypto-currency vs banks or cable TVs vs cellphones or Steam engines vs automated vehicles as well as other already affected industries. The bottom line here is that the internet has come to stay and change lives. Following recent statistics, regardless of the fact that even in this present age many are still to know of the ongoing change around the globe, close to 18billion devices are connected to the internet. Let's put it into perspective that these mega forces of disruption backed up by the internet are not yet fully around. In 2050, for example, would we have an entirely internet-based economy? One could only foresee in this age of artificial intelligence and the likes, that in coming years, humans would be able to re-doctor their forms to become half humans, half machines, leading to either extremes of technological Utopia or Hades.

## Why Is There Technological Disruptions?

The world is brimming with new guidelines and they do require transformational approaches on the dimension of individuals, and procedures. There are concerns to regulate the internet of things in the US and different nations. They want to have control of the blockchain technology by making it centralized, just as the European Union countries want to regulate privacy laws. All over the globe renewable energy and smart technologies are driving the plan in brilliant structures and savvy urban communities and championing the need for cities to become smarter and more connected. The need for connection and increasing complexity, which requires significant research, has led to many breakthrough discoveries in science, technology, business, and global economy. Disruptions are constantly about clients, workers, markets, competition, policies, and world issues. For instance, the change of information from paper into computerized data came from the need to increase productivity and the capacity for storing data in more secured platforms. Technology Innovation has greatly influenced the need for change in people, processes, systems and places. Technologies with clear disruption potential such as Social media, Mobile, 5G, WiFi 6/AX, Quantum Computing, Digital Twin, Autonomous Machines (Drone, Vehicle, etc.), Cybersecurity (privacy, availability, etc. discussed further in chapter 19), Internet of Things, Automation, the new reality (Augmented Reality, Virtual Reality, and Mixed Reality), Blockchain and 3D printing have significantly transformed old patterns and processes. The world is now a global space where billions of people on earth can connect. The most disruptive potential occurs when technology is modified with new applications. This can serve a mass market at an affordable price. The fundamental principle in disruption is the concept of empowerment and improvement which is introduced to several industries. Human needs and changes in customer demands can give rise to major innovative disruptions. Rapidly changing environments are where the speed of change touches upon a myriad of phenomena, ranging from the acceleration of technological innovation and disruptions challenging the status quo of common business models to the need for speed in dealing with changing customer demand across the value and supply chains. Technological disruptions are always about the needs of customers, workers, markets, competitors, health concerns, governments, security, and aging population stakeholders. Speed and accuracy have led to the disruptive effect.

## Prior to the Internet

Twenty years back, a business would open a retail store, place promotions in the neighborhood or newspaper, join a nearby systems administration association, and trust the nearby customers to refer them to other people. Such changed with the origin of the Internet. A business is never again reliant on its neighborhood customer base for its survival. It presently has a worldwide audience for its products. The Internet has changed a business customer base; however, it all depends on how a business speaks with its workers and finds and deals within the business ecosystem.

## AFTER THE INTERNET

**Capacity to Communicate**: A business' capacity to speak with its workers, customers, and partners changed significantly when the Internet introduced various mediums of connecting with customers, partners, employees, and all stakeholders of a business. Email and texting have changed the ways of business communication.

**Promoting to a Wider Area**: With the appearance of Internet promoting, a business must remain side by side with the necessities of its customers. The rivalry is never again restricted. A business presently has rivalry everywhere throughout the world. It is basic that a business understands what its customers need and delivers it. Reviews, surveys, feedback, and remarks can be integrated into an organization's website or social media platform to meet customers' needs better.

**Digital Advertising**: Considering the cost of advertising for companies, it sometimes stretches out the budget of some companies. Search engine optimization (SEO), enables a business to have a nearness on the internet and achieve a huge number of potential customers using the best keywords in the content of the digital adverts.

**Partnering with other Businesses**: Working with different organizations and experts is improved with the utilization of the Internet. Web classes, additionally called online classes, make teaming up on activities with individuals everywhere throughout the world as simple as signing onto a site.

**Internet Research**: Organizations utilize the Internet to look into new product ideas and new techniques for making products, and analyzing data. A business can likewise explore the challenge to perceive what products or services are advertised when an organization is hoping to venture into a specific area. The Internet can be utilized to inquire about the population, trends, needs and best seller products without having to go to the streets to make enquiries.

**Books**: Adolescents and children will never again prefer to utilize physical books. They now think of them as massive and awkward, however, that was not the case before. Physical books were the main source of picking up data or information on a particular subject. However, at this point with digital books, it is simpler to acquire data and learning. It has destroyed the inconvenience of visiting a book shop. Google is the best educator as it has the answer for all inquiries, data, and information needed at any time.

**Communication**: Like it or not, because of the Internet, communication today no longer has any restriction. Our brains have begun to work in unexpected ways; we prefer to send messages online instead of meeting face to face. To think about an individual, we check their Facebook or WhatsApp status or Twitter account instead of reaching them face to face. These days' individuals think of it as time wastage when they are approached to meet somebody face to face; rather, they would prefer video talk, remotely coordinating and other online methods of communication. The Internet overall has made communication simple.

**Online Shopping**: There is no need to visit a store anymore. Companies like Amazon and eBay have made the shopping experience easier and faster. There is no need to join queues for the purpose of shopping or the act of window shopping. You would simply be able to visit the online destinations' search at the best cost and request things that will be delivered to your doorstep. There is no longer any need to wait in queues when someone can deliver them directly to your doorstep.

**Travel**: Today, you can know everything about a spot before you reach there. You don't need to trust a travel guide to inform you of your favorite travel destination. There is a number of applications and GPS that will do just that for you. You can now analyze costs and book a taxi, flight tickets, transport, or train tickets effectively without waiting in lines. There is no longer any need for paper maps or to trust that a taxi would take you there.

**Relationships**: To know somebody, you don't need to meet him or her face to face. Presently, you can make friends on the internet, and meet the love of your life by essentially utilizing social applications. All anyone needs to know is that this is possible by means of a social application. Presently people don't depend on others, they depend on the reviews or remarks left on their profile.

**Healthcare services:** The Internet has made everybody their very own health specialist. Simply enter the symptoms you are feeling, and you could possibly have the right prescriptions recommended without visiting a doctor. Technology has progressed, however, not to the degree that you can depend on any and everything that you read on the internet. Particularly with regards to your wellbeing you shouldn't. Knowing things and picking up information is great; however, depending on them aimlessly without the counseling of a genuine doctor isn't great. Information will save you from being tricked yet won't treat you like a doctor can.

**Business**: To be profitable, is it important to go through eight hours in the office? Before, this inquiry sounded bad as the appropriate response was obvious. Yet, today things are distinctive. Internet has transformed them with online devices and steady network bosses, workers, both private and government, can keep in contact. Presently you can work from anyplace instead of adhering to the seat a desk area gets to your work records from anyplace and allows you to be profitable and progressively inventive.

## THE TECHNOLOGY SUPERGIANT TRENDS

Technology and digital trends will incorporate interconnected people, robots, gadgets, culture, content, health, scientists, researchers, engineers, programmers, and policymakers who seek digital transformation to find innovation patterns that would drive futuristic change and advancement. Innovation is presently advancing at such a fast pace, that yearly forecasts of patterns can appear to be outdated before they even go live. As innovation develops, it empowers considerably quicker change and advancement, causing the exponentially increasing speed of the rate of progress. Driving edge innovations, new interfaces are moving from consoles to touchscreens, and voice directions, rapidly changing the manner in which we connect with machines, information, and one another. The progression of significant innovation changes have started with the progress of centralized servers to personal computers and proceeded with the development of the web and portable devices. At each stage, the manner by which we interface with technology has become increasingly relevant and omnipresent. Think about the movement from console to mouse to touchscreen to voice and the subsequent changes to the manner we control onscreen information.

## ARTIFICIAL INTELLIGENCE TECHNOLOGY TRENDS

The manipulation of machines that work and respond like humans gave rise to the super trend known as artificial intelligence (AI). It is a hot trend in computer science which stresses the need to create highly intelligent machines that can think and make decisions just like humans can. A portion of the exercises computers with artificial intelligence can carry out includes man-made reasoning, speech recognition, learning, planning, and problem-solving. This software process has created savvy machines. Man-made brainpower is exceptionally specialized and concentrated knowledge, reasoning, problem understanding, perception, learning, planning, and the ability to control and move objects have enabled machines to act like humans by being able to act and respond like people having adequate data to identify with the world. Thinking and critical thinking power in machines is a gradual process of learning that would keep the world running by Artificial Learning. Alexa, Siri, Cortana and the various numbers of remote helpers make our lives a lot quicker. Systematic AI has qualities that are predictable with psychological insight, producing a subjective portrayal of the world and utilizing learning dependent on past experience to educate future choices. Human-roused AI has components from psychological and passionate knowledge, understanding human feelings, notwithstanding intellectual components, and considering them in their basic leadership. Refined AI demonstrates attributes of a wide range of capabilities (i.e., psychological, passionate, and social knowledge), can act naturally cognizant and is mindful in collaborations with others. In 1936, British mathematician Alan Turing applied his speculations to demonstrate that a figuring machine known as a Turing machine would be fit for executing intellectual procedures if it could be separated into various, singular advances and spoken to by a calculation. In doing so, he established the framework for what we call artificial intelligence today. In 1956, software engineer John McCarthy proposed to call this computerized reasoning. The world's first AI program, 'Rationale Theorist' figured out how to demonstrate a few dozen scientific hypotheses and information. With many years of research, artificial intelligence is still in its earliest stages. It needs to be increasingly solid and secure against control before it may be utilized in delicate zones like driving and medicine.

Machine learning enables computerized reasoning where systems learn and get better from experience capacity, making it possible for the machine to program itself. It is a subdivision of artificial intelligence which allows machines to develop, get information and use it learn how to enhance learning without any interference. The essential point is to permit the machine to adapt naturally without human input or help and alter activities as needs be. Machine learning calculations are regularly sorted as directed or unsupervised. Unsupervised machine learning calculations are utilized when the data used to prepare is neither grouped nor named. Unsupervised learning thinks about how frameworks can induce a capacity to depict a concealed structure from unlabeled information. The fundamental reason for machine learning is to analyze calculations that can get information and utilize facts to get desired results. Machine learning stages are among big business innovations' most focused domains, with most real sellers, including Amazon, Google, Microsoft, IBM, and others, hustling to sign clients up for different benefits of their machine learning platforms. Machine learning is getting systems or enabled devices to learn and act like people do, and improve their learning after some time in independent design, by sustaining information and data gathered. The early history of machine learning dates back to the primary instance of neural systems which was in 1943, when neurophysiologist Warren McCulloch and mathematician Walter Pitts composed a paper about neurons and how they work. They chose to make a model of this utilizing an electrical circuit, and along these lines, the neural system was conceived.

Deep learning is an AI procedure that encourages systems to do what works out easily for people to learn by model. Deep learning is a key innovation behind driverless autos, empowering them to perceive a stop sign, or to recognize a person on foot from a given distance. Systems figure out how to perform undertakings legitimately from pictures, content, or sound. Deep learning is a part of man-made consciousness (AI) that is concerned with imitating the learning approach that individuals use to increase specific sorts of information. Deep learning can allow systems to translate different languages and personalize shopping needs. It can be used in the health sector to detect sickness and customize a treatment that suits patients' needs and performs facial recognition. In 2012, Google Brain discharged the consequences of an irregular free-lively venture called the Cat Experiment which investigated the challenges of unsupervised learning. Profound learning sends regulated realizing, which implies the convolutional neural net is prepared utilizing named information like the pictures from ImageNet. This analysis utilized a neural net which was spread over more than 1,000 computers where ten million unlabeled pictures were taken arbitrarily from YouTube, as contributions to the preparation programming. From that year onwards, unsupervised learning remained a critical objective in the field of deep learning. Artificial intelligence is further discussed in chapter 5 below.

## Mega Trends in Cloud Computing

Cloud computing conveys hosted services and facilitated benefits over the Internet such as Infrastructure-as-a-Service (IaaS), Platform-as-a-Service (PaaS) and Software-as-a-Service (SaaS). Rapid developments in virtualization and appropriated processing, as well as improved access to the rapid Internet, have quickened enthusiasm for cloud computing. A cloud can be private or open. An open cloud provides services to anybody on the Internet who is willing to pay. Right now, Amazon Web Services is the biggest open cloud supplier. Private or open, the objective of cloud computing is to give simple, versatile access to computerized assets and IT administrations. In the open cloud model, an outsider cloud specialist organization conveys the cloud administration over the web. Open cloud administrations are sold on interest, on an hourly basis, where clients pay for the CPU cycles, stockpiling or transfer speed they expend. Leading cloud specialist organizations include Amazon Web Services (AWS), Microsoft Azure, IBM, and Google Cloud Platform. Security remains an essential concern for organizations considering cloud computing, particularly open cloud selection. An open cloud is a multi-occupant condition. This condition requests overflowing confinement between legitimate process assets. In the meantime, access to open distributed storage and register assets is watched by record login certifications. Cloud computing goes back to the 1950s when substantial scale centralized servers were made accessible to schools and organizations which led to the introduction of server rooms since the room would commonly just have the capacity to hold a solitary centralized server. As the expenses of server equipment gradually descended, more clients could afford to buy their own committed servers.

Edge computing is tied in with putting time-sensitive activities closer to where they will have an effect. Information gathering and investigation is assigned to where it can work most productively, instead of where it is most helpful for engineers or administrators. Edge Computing is handling of information at the edge of a system by handling information at the client end as opposed to being prepared in a neighborhood or virtual server. A perfect example of edge computing is an automated teller machine (ATM). The edge analysis gives information that can be utilized to take actions like shutting down a broken ATM and depleting it of cash in its apportioning unit. The internet of things is the most widely recognized use case for edge figuring, the internet of things is about the gathering of information from topographically scattered zones utilizing edge sensors. These sensors are associated with utilizing an information arrangement that regularly uses WAN advances. All gathered sensor information is transported to a focal storehouse where it is joined, and the information is handled on the whole. Imagine a scenario in which each internet of things sensor just needs to process the information it gathers and send results when certain prerequisites are met. By drawing processes nearer to the source of the information, the inertness or correspondences of data transfer capacity engaged with the roundtrip to the cloud get diminished.

Quantum computing could stimulate the advancement of new leaps in science, medicine and other fields that could save and protect human lives. Quantum Computing is a whole new paradigm shift to the entire way we compute and program. When Artificial Intelligence algorithms take advantage of Quantum Computing, it could accelerate many areas of research, i.e. analyzes data and detect diseases sooner and faster. Quantum Computing, when reality, is expected to create major impact to many industry, selections of materials to make increasingly effective gadgets and structures, money related systems to live well in retirement, and solutions to rapidly coordinate assets in cases of emergency. Quantum Computing is used today by highly developed countries in technology such as the United States and China. For issues over a specific size and multifaceted nature, we do not have enough computational power on Earth to handle them, but to stand an opportunity at taking care of a portion of these issues, we need another sort of processing. General Quantum Computers influence the quantum mechanical wonders of superposition and snare to make expresses that scale exponentially with a number of qubits, or quantum bits. There are a couple of various approaches to make a qubit. One strategy utilizes superconductivity to make and keep up a quantum state. To work with these superconducting qubits for broadened timeframes, they should be kept extremely cold. Quantum computing works at temperatures near supreme zero, colder than the normal room temperature. Working with Nano-scale segments at temperatures colder than intergalactic space, quantum computing can possibly settle a portion of the world's hardest difficulties, taking just days or hours to take care of issues that would take billions of years;

quantum computing would empower new disclosures in the territories of social insurance, clean energy, ecological frameworks, smart materials, and several other industries.

## Internet Megatrends

Accessibility of the Internet was once constrained; today the acceptance and the growth of internet users across the globe have rapidly grown. In 1995, just 0.04 percent of the total populace had internet access with the majority of users in the United States. By the beginning of the 21st century, numerous purchasers in advanced countries utilized quicker broadband services, and by 2014, 41 percent of the total populace was able to connect on the internet. Broadband was practically omnipresent around the world and worldwide as normal speed surpassed one megabit for each second. Early investigation into Internet imbalances between the advanced nations and the developing nations brought about great concern because the goal of the World Wide Web was to connect the whole world digitally. Practically 4.4 billion people use the internet regularly as of April 2019, including 58 percent of the worldwide populace. China, India and the United States rank ahead every single other nation regarding the highest internet usage. One reason numerous individuals are not signing on the internet yet is due to the fact of constraint and lack of accessibility – 31% of the worldwide populace do not have 3G inclusion, while 15% have no power. In sub-Saharan Africa somewhere in the range of 600 million individuals (right around 66% of the district's populace) do not have normal power, and this applies to about a fourth of individuals living in South Asia. The expense of gadgets and network is another factor keeping numerous individuals from getting access to the web, particularly the 13% of the total populace living underneath the destitution line. Broadband is moderate for 100% of the populace in only 29 nations. The worldwide online infiltration rate is 57 percent, with North America and Northern Europe both positioning first with a 95 percent web entrance rate among the populace.

It has been a couple of years since the United Nations Human Rights Council perceived web access as a fundamental human right. Cell phones have turned into the most mainstream approach to get on the web and we depend on two innovations to associate: Mobile systems and Wi-Fi. The two advances are always developing and improving. The possibility of 5G versus Wi-Fi is deceitful on the grounds that what we truly need is both. 5G is the umbrella term for the fifth era of portable mobile technology, and it envelops a variety of components. Cell or versatile systems depend on authorized range groups, which are sold to the best bidder, such as Verizon or AT&T. To take off inclusion they need stations fit for conveying a sufficient capacity that can serve various individuals in thousands who live in urban zones without any delay. To recover their returns on investments, they depend on having subscription packages for their customer base. The possibility of download speeds somewhere in the range of 10Gbps, and transfer speed of only 1 millisecond has gotten individuals amped up for 5G. However, actually, we don't regularly go anyplace close to the hypothetical top velocities. The genuine speed of 5G association will rely upon numerous components including location, the internet service provider (ISP), the number of users, and the model of the gadget being used.

Wi-Fi has customarily been befuddling as far as the naming shows for models. It went from 802.11b to 802.11a, 802.11g, 802.11n, and after that 802.11ac, yet fortunately, the Wi-Fi Alliance has acknowledged the requirement for something less confusing thus the following standard, 802.11ax will be called Wi-Fi 6. This easier naming show is likewise being retrofitted, so 802.11ac will move toward becoming Wi-Fi 5, etc. The new Wi-Fi 6 standard should offer speeds no less than multiple times more prominent than current Wi-Fi solutions, yet it will likewise acquire enhancements productivity intended to adapt to the developing number of gadgets in the normal home that associates with the web. Much the same as 5G, Wi-Fi 6 will supplement, not supplant, existing Wi-Fi models. According to Cisco Systems, the Wi-Fi 6 solution, is a new promising technology that will enable the NexGen mobile experience. It will offer faster speeds and quality for immersive experiences, and will be the mobile platform to facilitate more IOT devices and much higher density environments.


## DIGITAL TWIN MEGATREND

A computerized twin is an advanced portrayal of a physical item or framework. The innovation behind digital twins has extended to incorporate huge things, such as structures, industrial facilities, and even urban communities, and some have said individuals and procedures can have advanced twins. The concept of digital twin came from national aeronautics and space administration (NASA) to help them create replicas of early space cases, that would reflect and analyze issues in a circle, based on computerized simulations. A digital twin is a computer program that takes genuine information about a physical environment and develops predictions of how that physical framework will be influenced by those data sources. Drones, trains, driverless cars, and turbines can be planned and tried carefully on digital twin ecosystems before being physically created. These computerized twins could likewise be utilized to help with upkeep activities. Automotive computerized twins are made conceivable on the grounds that autos are as of now fitted with telemetry sensors, yet refining the innovation will turn out to be increasingly significant as progressively self-ruling vehicles hit the street. Digital twins can be utilized to anticipate distinctive results depending on factor information. Computerized twins can regularly advance an internet of things organization for most extreme productivity, just as help planners make sense of where things ought to go or how they work before they are physically conveyed.


## AUTONOMOUS MACHINES MEGATREND

Autonomy is the capacity to settle on your own choices. In people, autonomy enables us to do certain tasks like strolling, talking, waving, opening entryways, and changing lights. In robots, autonomy is actually the same when compared to human behavior. Autonomous robots require a motion, mapping, navigational calculations, and natural detecting innovation to effectively explore environments and relate with humans. Drones can be remotely controlled or can fly independently through programming-controlled flight designs in their installed frameworks working related to locally available sensors and GPS. Cutting edge automatons and robots can be more intelligent, lighter and quicker than their previous models. The pitfall in autonomous machines is the constraint of the machines not being able to know the actions to take in cases of emergencies or situations they have not been designed for.

# THE INTERNET OF THINGS

The internet of things is a connection arrangement of interrelated processing gadgets, mechanical and advanced machines, people and communities that are given special unique identifiers (UIDs) and the capacity to exchange information over a system without expecting human-to-human or human-to-PC communication. The Internet of Things (IoT) works by essentially interfacing any gadget with an on and off change to the Internet or to one another. This incorporates everything from smartphones, coffee makers, clothes washers, earphones, lights, wearable gadgets and nearly whatever else you can consider. The Internet of things can be connected to things like transportation systems by creating a smart city that would enhance efficiency and productivity. Security is a major issue that needs to be looked into. With billions of gadgets being associated together, individuals need to ensure their data remains secure. At that point, we have the issue of security and information sharing. The history of the internet of things can be traced back to Kevin Ashton, who first referenced the internet of things in an introduction he made to Procter and Gamble (P&G) in 1999 on needing to bring radio recurrence ID

(RFID) to the consideration of P&G's senior administration. Ashton called his introduction "Internet of Things" to consolidate the cool new pattern of 1999. Despite the fact that Ashton's was the principal notice of the web of things, associated gadgets have been around since the 1970s.The principal web apparatus, was the Coca-Cola machine at Carnegie Mellon University in the mid-1980s which utilized the internet of things by the then software engineers when they needed to check the status of the machine and decide if there would be a very cold drink or hot drink available in the machine before they arrive at the location of the machine. Internet of things developed from machine-to-machine (M2M) correspondence to machines interfacing with one another by means of a system without human connection. M2M alludes to associating a gadget to the cloud, overseeing it and gathering information. The idea of the internet of things became a phenomenon in 2010. Countries like China have promised to make it a vital plan five years from now.

## THE NEW REALITY

The digital world and the real world continue to experience transformations. Augmented reality (AR) is rapidly developing in bringing components of the virtual world, into our genuine world, therefore improving the things we see, hear, and feel. Augmented Reality brings illustrations, sounds, and simulations into real characters in the real world. Virtual reality can help us manipulate our physical world to whatever we might want it to be, and also help us produce some truly stunning things. Virtual reality places humans inside an encounter and readies them to interface with 3D universes. One day we will almost certainly put on a pair of glasses and be transported from our worries into our wonderland; one day these glasses, which are as light as a couple of Ray-Ban's, will give us a chance to swim the profundities of the Great Barrier Reef while staying dry in our family room. One day Virtual Reality and Mixed Reality will be key to our work, correspondence, excitement, and exploration. Mixed reality will unite the physical and digital world with just a single device. Blockchain will enable computerized data to be disseminated without being duplicated, while 3D printing would bring an imaginative innovation that would make humans be able to print physical items from a computerized model. It would now be easier to print items from home rather than going out to purchase them.

## What is the Internet of Things?

The Internet of Things is incredible, yet it's insufficient. To understand the genuine capability of the IoT, we should think all the more extensively, and consider the IoT as a segment of a whole. While IoT is explicitly about empowering non-processing gadgets to share information through the Internet, the Internet of Everything is a lot more extensive. IoT is a basic part; however, the Internet of Everything is tied in with following up on all the data accumulated from the four essential mainstays of the Internet of Everything: Data, Things, Processes, and People. The Internet, obviously, was created as a way to share and transport information. Information will continue to be at the center of each web innovation, stage, and worldview. Remember that strategies for correspondence, for example, voice, content, and video are largely still just information, even with no sort of innovation included. So, what is unique? The information has, for the most part, been produced, stored away, at the point transmitted, starting with one spot, then onto the next; or communicated from one point to another focuses (spilling video, communicating radio, enabling satellite transmissions and so forth.). As referenced previously, the IoT is a key mainstay of the Internet of Everything. The things are the gadgets huge and little, perplexing and straightforward, that both feed the information into the Internet of Everything, and much of the time, follow up on the information moving through the Internet of Everything. Robots, clever traffic lights, home machines… all conceivably fit inside the Internet of Everything as IoT empowers gadgets. The Internet of Everything is about the circumstance. It is tied in with making life simpler, increasingly agreeable, and making new potential outcomes and alternatives that did not exist previously. 20 years back, Facebook, WhatsApp and cell phones, were scarcely possible, and GPS was an innovation utilized by the military and experts. Digital photography was as yet the undesirable stepchild to film. We know the Internet of Everything is here and developing, yet no one is insightful enough to imagine the absolute potential yet to be figured out.

Billions of physical gadgets around the globe that are connected to the internet can gather and share information through the internet of things with good processors and wireless networks. It's conceivable to turn anything, from a pill to a plane to a self-driving vehicle into part of the IoT, empowering them to convey ongoing information without a person included, successfully combining the digital and physical universes. There are now more associated things than individuals on the planet. Around 8.4 billion IoT gadgets were being used in 2017, up 31 percent from 2016, and this will probably reach 20.4 billion by 2020. All together spending on IoT will reach nearly $2tn in 2017, with 66% of those gadgets found in China, North America, and Western Europe.

**Internet of Things applications at this moment**

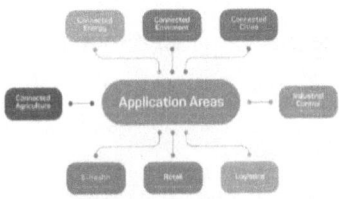

**Smart homes**: The IoT Analytics organization database for Smart Home incorporates 256 organizations and new businesses. Large number of organizations are dynamic, when it comes to smart home soltions, than some other applications in the field of IoT. The aggregate sum of subsidizing for Smart Home new companies right now surpasses $2.5bn. This rundown incorporates unmistakable startup names, for example, Nest or AlertMe just as various global enterprises like Philips, Haier, or Belkin.

**Wearables**: Wearables remains an intriguing issue as well. There are a lot of other wearable developments to be amped up for: like the Sony Smart B Trainer, the Myo motion control, or LookSee arm ornament. Of all the IoT new businesses, wearables creator Jawbone is presumably the one with the greatest subsidizing to date. It remains at the greater part a billion dollars!

**Smart City**: Smart city traverses a wide range of utilization cases, from traffic, to water circulation, urban security, and climate checks. Its prevalence is energized by the way that many Smart City plans guarantee to ease the standard of living for people in urban areas. IoT plans in the zone of Smart City to tackle traffic blockage issues, air pollution, and contamination and help make urban communities more secure.

**Augmented Reality**: Augmented reality has been utilized in manufacturing for quite a while. AR has been used to overlay digital graphics created sounds onto the physical environment using real-time sensor information. For instance, you can overlay the computerized portrayal of a bit of hardware onto the manufacturing floor to see where it would fit best. Hardware OEMs can utilize augmented reality (AR) as a presentation of constant information layered on a live video to further improve their field administration capacities and contributions.

**Virtual Reality**: Virtual reality innovation, spearheaded by the gaming business, gives an increasingly vivid, computer stimulated experience. With VR, everything is digitized and can be controlled, empowering people to communicate as though they were directly there. VR depends on headsets and goggles to make this present reality fantasy and regularly coordinates sound and vibrations. The main headset, Microsoft HoloLens, draws in people with virtual encounters that can fuse visual, sound, speeding up, and balance components. The early use of VR in the industry was for operator training simulators (OTS). For some unsafe enterprises, similar to the nuclear business, OTS preparing is ordered, since it has demonstrated to be increasingly viable in preparing activities to react rapidly and adequately to abnormal circumstances.

**Digital Twin and Simulation**: Various organizations characterize the digital twin somewhat diversely relying on their separate innovation stack. As a rule, the digital twin is a virtual or advanced portrayal of physical or real-world assets that incorporates demonstrating conduct and real-time analytics and machine learning. Digital twin representation is a mix of virtual, augmented, and mixed, extended reality that reproduces this present reality in the world. The software digitalizes and mimics a real procedure, machine, or thing utilizing real-time sensor information and models that empower people to cooperate in a virtual way.

## The Extended Reality of the Future

In the future, Virtual Reality (VR), Augmented Reality (AR), Digital Twin (DT), and Mixed Reality (MR) will cooperate and disturb how we work together. Removing the divide between the Physical & Virtual worlds, is a major transformation in the behavior of people and how they interact with reality & virtualization. Not exclusively will the physical world be joined with the virtual world, yet it will incorporate substance and organic characteristics of materials and things. Smartphones, versatile VR headsets, and AR glasses will combine into a single expanded reality; wearables will replace every other screen in your life. These wearables will give vivid, subjective, and connected encounters.

Expanded reality will likewise assume a bigger job in manufacturing from product plan and prototyping to helping administrators, and different specialists take care of procedure issues, keep up a sheltered situation, and look after equipment.

There will be difficulties in utilizing these innovations, for example, cost, process limit, stockpiling, and execution. Extended reality requires gigantic volumes of information and regularly at quick speeds. As of now, making an application like this is generally costly, even with the present industrial internet of things stages, related administrations, and AR application improvement programming. Databases should coordinate, contextualize, and store information in an information store that empowers information, investigation, and different applications. New sorts of cybersecurity should be set up to do this safely. Better registering execution, better battery innovation, incorporated materials and science, and more institutionalization will further progress an extended reality. Consistent availability supplemented by 5G and gigabyte LTE innovation and propelled 3D representation will prompt extra utilization of expanded reality technologies. Extended reality will help robot-based manufacturing, remote services, and training, including preparing for perilous situations that are too dangerous to even think about training in physically, and train individuals to react to potential calamities that may never occur.

## The Predictive Digital Twin

Digital twin helps in visualizing objects in order to display information about the data gathered from the object with the use of sensors and digital technologies. A predictive digital twin goes above and beyond; it opens current-state information to drive resource and operational process performance. It is made utilizing uncommon predictive simulation software and can be utilized to assess future situations. Utilizing this digital twin helps decision makers test and comprehend the effect of various situations, thereby distinguishing from opportunities or dangers without causing any expense, and finding solutions utilizing dynamic models of real business and operational procedures.

## Some Use Cases of Digital Twin

**Effectiveness**: A wind farm can effectively use the digital twin to showcase wind patterns, wear on hardware and electrical yield. The data can be utilized to improve the proficiency of activities.

**Dangers**: The engine of a plane can be seen in flight including temperatures and weights on all the parts of the plane. The model can be utilized to identify dangers or safety risks before they happen.

**Assembling**: A digital twin of an air-conditioning unit can be used to recognize quality issues in the unit during assembling of parts in the manufacturing plant.

**Smart City**: A smart city platform gives an ongoing controllable model of a downtown area that incorporates underground structures, for example, water frameworks. Teams can investigate the foundation and administrations from a remote area.

**Testing**: A digital twin of a smartphone is utilized to envision the performance of the device under various conditions, such as when it overheats, drops, is smashed, put into water and other conditions to test for possible solutions or other ways to enhance the life cycle of the smartphone.

**Design**: 3D printer designers can consistently review a digital twin of a working printer to recognize configuration deformities, such as parts that run hot or materials that obstruct parts. A digital twin could help a surgeon effectively visualize a heart before working on it, thereby minimizing human error in surgical operations.

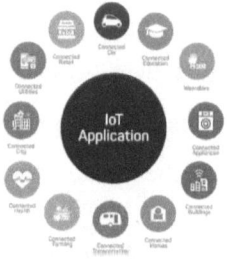

## Human Concerns in The Digital Transformation Era

The late Professor Stephen Hawking, Bill Gates and Elon Musk have frequently communicated concerns with regards to computerized reasoning (AI). They stressed over what will transpire as AI turns out to be progressively complex and more astute.

Cybersecurity is the act of ensuring that frameworks, systems and projects are fully secured from advanced assaults. Executing powerful cybersecurity measures is necessary today for the fact that there is a greater number of gadgets than individuals, and assailants are becoming increasingly creative in hacking devices daily, more so why digital security is paramount.

There is also the concern of automation by robots to perform undertakings that were recently done by humans. Digital transformation (fully discussed in the next chapter) is to help productivity and dependability, in most cases, technological advancements create unemployment for some individuals by replacing their jobs with robots or machines; most of the time.

All the mega forces can create real technological disruptions at remarkable speed. Artificial intelligence being one of the mega forces of disruption is a force on its own that some refer to as a revolution within the revolution 4.0. It's then very essential that we take a closer look at what artificial intelligence is all about and the magnitude of advancement it is bringing to the world we live in today. The next chapter discusses essential details about artificial intelligence and the human-machine.

# CHAPTER 4

## ARTIFICIAL INTELLIGENCE AND THE HUMACHINE

The concept of Artificial intelligence was first used by McCarthy in anticipation of a 1956 summer research project at Dartmouth College. The Society for the Study of Artificial Intelligence and the Simulation of Behavior (SSAISB or AISB), characterized AI at the time as "the science of making machines clever". This exploration task is currently generally viewed as the underlying impetus that anticipated AI into turning into its very own order. The field has made some amazing progress from that point forward and is relied upon to keep advancing at an exponential rate, yet what's to come is famously hard to anticipate. Luckily, the advancement of numerous innovations appears to pursue a typical long-term design. In this way, it is conceivable to pick up a moderately dependable thought of how AI will progress sooner rather than later by taking a look at how it has propelled over the years. This gives a premise from which estimations can be drawn about how AI will fit and shape social orders and advances of things to come. This field is getting popular, especially in the areas of optimization, automation, and information analytics; a lot of research has been carried out to advance the cause of artificial intelligence over the years.

Automation has, for some time, been utilized to decrease both work and material expenses while at the same time improving the nature of the finished result. As machines don't get exhausted or occupied, they can perform with more consistency than a conventional human workforce. In spite of the fact that this has been constrained to generally repetitive errands, as AI research progresses, there would be more unpredictable tasks for machines to handle. Without a doubt, human work will be altogether replaced by the workforce of robots in the next 125 years to come. Gradually, several organizations are putting resources into driverless cars due to the fact that their automated response times and consistent readiness, could improve general security. Also, streets with no human drivers will probably have fewer mishaps, contamination, and congestions. In fact, the trucking industry presently has various issues because of exhausted drivers, an issue that could be immediately illuminated by presenting automated trucks, which would not come with a lower cost of maintenance and risks. The American Trucker Association, as of now reports that there are over 3.5 million expert truck drivers in the USA; these workers would have their jobs compromised if driverless trucks became promptly accessible. Obviously, there is a dilemma between better wellbeing and proficiency against professional stability. One potential arrangement is to have AI assume the job of a partner. For instance, AI could be utilized to caution drivers when they are out of path, control crisis brakes, or even drive autonomously on motorways. Tesla has, as of late, declared designs to discharge a truck with these capacities in 2019. In this way, AI can be utilized to expand human execution during a particular task, by dealing with the components that people are not generally excellent at. Consistency and precision can be leveraged from automated machines. They can likewise process more calculations than humans, which makes them capable when managing a lot of data.

Without a doubt, if an AI system is given an assignment, with a set number of quantifiable parameters and objectives, it can produce endless solutions. The best of these solutions could possibly take after the procedure of evolution. Consequently, these are known as hereditary algorithms or generative structures. These strategies result in very improved yields. Airbus has utilized generative procedures to design new parts for its planes. The recently acquired ones were 45% lighter and utilized 95% less raw material during their production than the past ones. In view of the sheer volume of solutions being considered, AI regularly thinks of results that would be incredible to people. This makes it a superb tool for thinking of new answers for old issues. For instance, the construction business hasn't changed much since the 1940s and needs another methodology. As each structure is moderately one of a kind, it sets aside a ton of effort for people to plan them. This significantly lessens the business' effectiveness. Investigations showed that this issue could be illuminated by utilizing generative plans to deliver altered structures in large volumes. Access to some level of personalization would improve various fields, particularly those that, similar to the construction industry, haven't changed much in the previous century, for example, agriculture, real estate, and education.

Similarly, as with automated trucks, generative designs are probably going to be utilized to upgrade human execution as opposed to replace them. Such technical algorithms might be very complicated for designers who have no technical background. Subsequently, a major obstacle for this technology is making the procedures available enough for feasibility in order to affect future technology.

Machines have more capacity to deal with enormous data as compared to humans. AI-driven innovations will become increasingly noteworthy and progressively typical as we acquire more information. The number of gadgets that are associated with the internet is going to surpass 50 billion before the end of 2020, and these gadgets will not just be smartphones and computers but other home and office appliances. As basic gadgets and appliances would also be connected to the internet of things. This Internet of Things (IoT) gives an overflow of data which can be utilized to foresee things from market patterns to medical issues. However, current diagnostic methodologies don't scale up to the sheer volume of information produced by the IoT, and AI may be depended on to understand it. Here, AI takes on a warning job, consequently transferring predictions to people dependent on a consistent, and perhaps a live stream of data.

When we consider 2050 it appears as though it is ages from now and we envision a totally extraordinary world.

Yet as a general rule, it is only a long time from now, and we would already be able to understand what will be the reality at that point. We have a great deal of ecological, social issues, and it is high time we determine how innovation may settle them by 2050.

If the field of AI keeps moving in similar ways, one can expect that innovation will begin to expand and improve current human abilities by enhancing them with automated advisors and collaborators. Keep in mind that the issues of regular day to day existence are conceivably difficult for a single research field to cover. Artificial intelligence has just demonstrated its ability to work nearby people in various spaces, improving the exactness of cancer growth and disease identification, decreasing exhaustion and improving smoothness during a medical procedure and lessening contamination in farming. In this way, the working dynamic among people and AI is by all accounts moving from a master-slave relationship to a partnership relationship.

Artificial intelligence will keep on advancing such that it will empower new disruptive trends in business and ecosystems. AI will improve human life by empowering people to finish tasks more effectively and in less time. Toward this path, deep learning innovation is relied upon to assume a key job in analyzing complex patterns more than what AI can at present give. By and large, we ought to anticipate the accompanying patterns in the trend of AI:

Installed AI: AI will be embedded in gadgets in different settings, including vehicles, customer gadgets, web indexes, and that's only the tip of the iceberg. We can expect the vast majority of things to become gadgets that include the Internet of Things to utilize AI capabilities.

Artificial intelligence machines are becoming more proficient than humans: In areas of information processing, AI machines will be substantially more productive than individuals; however, in the area of creativity and human skills, AI still has a huge gap to bridge. Moreover, individuals will progressively prefer to get services from machines instead of people because of their proficiency and zero-blunder execution.

Everyday activities: Personal errand robots will multiply and begin helping individuals in completing an ever-increasing number of tasks. Artificial intelligence will help humans in their everyday tasks, for example, cooking, treatments, cleaning, installations, and other tasks humans may be too tired to carry out.

Self-driving vehicles and more secure transport: The approach of AI-empowered self-driving cars will lessen the check of mishaps fundamentally, saving a huge number of lives and prompting tremendous financial savings.

Increasing human capacities: AI gadgets and robots will later on give us superhuman capacities, by broadening our present abilities by enabling humans to be quicker at handling information, just as far as performing manual errands that are unrealistic these days.

A portion of our closest friends will be robots: Future robots are probably going to be our partners and assistants. To this end, future robots might be additionally ready to comprehend human behavior while understanding our feelings. As portrayed in the motion picture, "HER" this relationship probably doesn't sound engaging today but will give a profoundly new human-driven measurement to AI.

But then it is never conceivable to totally anticipate the eventual fate of technology and where it's going. We do have great proof that AI will eventually disrupt our regular day to day existence while improving the world, and we ought to plan to encounter this insurgency and understand all that it brings to the table.

## GROUPING OF ARTIFICIAL INTELLIGENCE

Artificial intelligence (AI) helps with the simulation of human insight formed by machines, particularly computer systems. Particular applications of AI include expert systems, speech recognition, and machine vision. These procedures incorporate learning (the securing of data and guidelines for utilizing the data), thinking (utilizing standards to achieve estimated or clear ends) and self-adjustment. Specific uses of AI include machine vision, expert system, and speech recognition.

Artificial intelligence can be classified as either weak or strong. Whereby weak AI, could be termed as limited AI, because it is structured and prepared for a specific task. A virtual personal assistant like Apple's Siri, is a type of weak AI. While strong AI, otherwise called artificial general intelligence, is an AI framework with cognitive human capacities. At the point when given a new assignment, a strong AI framework can discover an answer without human efforts.

Since programming, software and staffing costs for AI can be costly, vendors are incorporating AI parts in their standard offerings, just as well as access to Artificial Intelligence as a Service (AIaaS) platform. Artificial Intelligence as a Service enables people and organizations to try different things with AI for different business purposes and test numerous samples before making a decision on what they want to buy. Well, known AI cloud offerings include IBM Watson Assistant, Google AI services, Amazon AI services, and Microsoft Cognitive Services.

While AI tools present the scope of new usefulness for organizations, the utilization of artificial intelligence brings up ethical issues. This is on the grounds that profound learning algorithms, which support a significant number of the most exceptional AI tools, are just as smart as the information they are given during the training phase. Since a human chooses what information ought to be utilized for preparing an AI program, the potential for human inclination is inborn and must be observed intently.

Some industry experts have the perception that artificial intelligence is firmly connected to mainstream culture, thereby making the overall population have ridiculous apprehensions about artificial intelligence and impracticable assumptions regarding how it will change the work environment and life when all is said and done. Advertisers and researchers trust that the tag artificial intelligence, which has a progressively nonpartisan implication, will help people better understand that AI will basically improve products and services, and never replace those using the application.

## SEGMENTS OF AI

Artificial intelligence is fused into a wide range of technologies. Here are some models:

**Automation:** What makes a framework or procedure work naturally? For instance, robotic procedure automation (RPA) can be modified to perform high-volume, a repeatable task that people typically performed. RPA is not the same as IT automation in that it can adjust to evolving conditions.

There are three types of machine learning algorithms:

**Machine Learning:** The exploration of getting a computer to act without programming. Deep learning is a subset of machine learning that, in basic terms, can be thought of as the automation of predictive analysis. They fall under these AI algorithms like supervised learning: data sets are marked with the goal that pattern can be distinguished and used to name new data sets. Unsupervised learning: data sets are not marked and are arranged by likenesses or contrasts. Reinforcement learning, where data sets are not named but, in the wake of playing out an activity or a few activities, the AI framework is given criticism.

**Machine vision:** The exploration of enabling the computer to visualize, this innovation catches and dissects visual data utilizing a camera, similar to digital change and digital signal processing. It is regularly contrasted with human visual perception, yet machine vision isn't bound by science and can be programmed to see through walls, for instance. It is utilized in the scope of applications from proof of signature investigation.

Computer vision, which is centered on machine-based image processing, is frequently conflicted with machine vision.

**Natural language processing (NLP):** This is the process of handling of human and non-computer language by a computer program. One of the best-known instances of NLP is spam detection, which looks at the title and the content of an email and chooses if it's garbage or not. Current ways to deal with NLP depend on AI. Natural learning processing includes speech recognition, translation and sentiment analysis.

**Robotics:** This is a field of engineering that basically focuses on the manufacturing of robots that would be used to perform tasks that are hard for people to perform or perform reliably. They are used in assembly lines for the production of vehicles or by NASA to move huge objects in space. Scientists are utilizing machine learning to build robots that can interface in social settings.

**Self-driving cars:** This mix of image recognition, computer vision, and deep learning to assemble a robotized skill at steering a vehicle while remaining on a given path and maintaining a strategic distance from any form of obstruction is one of the innovations self-driving cars has.

## AI APPLICATIONS

Artificial intelligence has advanced into various regions.

**Artificial intelligence in Healthcare:** Here the focus is centered on the best ways to improve the health of patients with minimal cost. Organizations are applying AI to improve and diagnose diseases faster. The best-known services innovation in this area is IBM Watson. Natural language is easily understood and it is equipped for reacting to inquiries posed of it. The framework mines quiet information and other accessible information sources to shape a theory, which is at that point present with a certainty scoring outline. Other AI applications incorporate chatbots (a computer program utilized online to respond to questions and help clients, help timetable follow-up arrangements, and help patients through the procedure) as virtual health assistants in order to give medical feedback that would improve the healthcare sector.

**Artificial intelligence in business:** Robotic process automation is being connected to tasks that are repeated by people. Machine learning algorithms are being linked to CRM platforms and analytics to reveal the best possible way to serve clients. Chatbots have been fused into sites to give prompt customer service to clients. The automation of job positions can be turned into an idea among scholars and IT experts.

**Artificial intelligence in education:** A lot of tasks in education can be automated by artificial intelligence; tasks ranging from grading, reports to parents, school fees and several others, thereby giving educators more time to advance in their career. Artificial intelligence can evaluate students and adjust to their needs, helping them work at their very own pace. Artificial intelligence educators can give extra help to students, by making sure they remain on track. Artificial intelligence could change where and how students learn or change some teachers that are not delivering quality education to students.

**Artificial intelligence in the financial sector:** In the area of finance applications such as Turbo Tax or Mint, artificial intelligence would eventually disrupt financial systems. This application generally gathers the information of its users and gives the best financial decisions to them. For example, IBM Watson, has been connected to real estate, whereby buyers could purchase their choice of houses with ease. Today, these financial applications perform lots of trading activities on Wall Street.

**Artificial intelligence in law:** This sector has a lot of documentation such as filing, cases, reports and several others for lawyers and clerks which is truly overwhelming for people to handle. The automation of this process would make this sector more productive and efficient with time management.

**Artificial intelligence in manufacturing:** This is a region that has been at the cutting edge of joining robots into the workforce. Industrial robots are used to perform separate tasks from the tasks of human workers, but technology advancement changed the whole process.

# THE IMPACT OF AI ON HEALTHCARE

AI systems can help physicians in a lot of ways. Since they are armed with a lot of information, they can help in clinical decision making. Diagnostic errors and therapeutic errors can also be minimized safely. Besides, AI systems have access to immense volumes of data enabling them to make predictions about health by extracting useful information about potential health risks. Its effectiveness is beyond doubt. And here is why:

- Hospital error has been one of the highest causes of patients' death. Such errors can definitely be prevented by Artificial Intelligence.
- Statistically, medical errors which could easily have been prevented by AI are responsible for the deaths of about 440,000 Americans each year.

In the next 5 years, the AI health market will grow by more than 10 percent. From early detection to improved diagnosis, AI is positively contributing to the advancement of humanity and society. In some areas, it is already being used, and there are areas where we can see the potential of AI in the nearest future. AI has started delivering high value in specialist areas such as pharma, pathology and radiology. Chronic health conditions are expected to be the area that benefits the most from AI's employment in the field such as cancer and heart diseases.

## ETHICAL IMPLICATIONS OF ARTIFICIAL INTELLIGENCE

AI poses a whole new set of ethical challenges and implications for business leaders whose use of mechanization may have profound effects on the workforce and society. In the haste in adopting the rapidly developing technology, organizations may run the risk of overlooking potential ethical implications which could produce unwelcome or unsavory results, especially in artificial intelligence (AI) systems that encompass machine learning. Machine learning is a subset of AI in which computer systems are taught to learn on their own. Algorithms are used to enable the computer to analyze data to detect patterns and gain knowledge or abilities without having to be specifically programmed. It is this type of AI that makes voice-enabled assistants such as the Google Assistant, possible.

In the financial or banking sector, the potential applications of AI include real-time auditing and analysis of company financials. Data is the raw material, so to say, that powers machine learning (further discussed in chapter 11). But what happens if the data given to the machine is flawed, or the algorithm that guides the learning isn't properly configured to assess the data it's being fed? Things would obviously go very wrong quickly. Research has identified a number of challenges facing business leaders. They include questions like,

What degree of control can organizations retain over our machines' decision-making processes?

How can we ensure that the systems act in line with the organization's core values?

Since biased algorithms can lead to a discriminatory impact, how can we ensure fairness and accuracy?

The report proposes a framework that outlines ten core values and principles for the use of AI in business settings. They are intended to minimize the risks of ethical lapses in event of improper use of AI technologies.

- Accuracy
- Respect of privacy
- Learning
- Transparency
- Interpretability
- Fairness
- Control
- Impact
- Integrity
- Accountability

## WHY IS ARTIFICIAL INTELLIGENCE IMPORTANT?

AI has been able to automate the process of repetitive learning and data discovery. But AI is not the same with hardware-driven, robotic automation. Rather than automating manual tasks, AI performs frequent, high-volume, computerized tasks without any form of weakness. For this type of automation, human inquiry is very much needed to ask the right questions.

Artificial intelligence adds knowledge to existing products. By and large, AI won't be sold as an individual application. But products would be improved with the use of AI applications, just as Siri was added to the latest Apple products. Bots, conversational platforms, smart machines, and automation can be joined with a lot of information to improve several technologies used in homes and working environments, from investment analysis to security intelligence.

Artificial intelligence constantly adjusts through dynamic learning algorithms to give the information a chance to do the programming. Computer-based intelligence discovers structure and regularities in information with the goal that the calculation gets an expertise opinion: The algorithms turn into a classifier or an indicator. Thus, similarly as the algorithms can show themselves how to play chess, they can show themselves what product to suggest next on the internet. What's more, the models adjust when given new information. Back propagation is an AI procedure that enables the model to alter, through preparing and including data, when the primary answer isn't exactly correct.

Artificial intelligence breaks down more and more profound data by utilizing neural networks that have many concealed layers. Building a fraud detection framework with five concealed layers was practically inconceivable a couple of years back. Every one of that has changed with unimaginable big data and computer power. Profound deep learning models are highly required in order to enable learning from different data sources; the more the information the better the result.

Artificial intelligence accomplishes amazing exactness through deep neural networks, which was previously termed unachievable. For instance, your communication with Alexa, Google Search, and Google Photos is altogether founded on deep learning, and they continue to get progressively precise the more we use them. In the medical field, AI systems from deep learning, image and object recognition can presently be used in the discovery of cancer cells on MRIs with similar precision to a trained radiologist.

Artificial intelligence makes good use of data. At the point when algorithms are self-learning, the information itself can end up becoming an intellectual property. The appropriate responses are in the data; you simply need to apply AI to get them out. Since the job of the data is currently more significant than at any time in recent history, it can be of great benefit in a very competitive industry, because only the best data would emerge as the winner when similar techniques are applied.

The forces stirring innovation and technology are wide and far reaching. Having discussed the front-liners of the mega forces of disruption and Artificial intelligence, this raises the need to address how we can easily navigate the new complex world that these mega forces are creating. We need to address how to easily navigate this new complex world. With every revolution comes a level of complexity and only those that are able to find their way around the complexity will be able to maximize their potential with the evolution of digital transformation. In the next chapter we will be looking at how we can navigate through the complexities created by the digital revolution.

# CHAPTER 5

## NAVIGATING THE NEW COMPLEX WORLD

Math is one of the most amazing subjects in all studies. It is an intriguing knowledge point to consider to a great extent, since it works uniquely in contrast to numerous different regions of human thinking. In arithmetic, the possibility that the expansion of two positive whole numbers can only ever lead to a positive whole number is proverbial. Regardless of what positive whole numbers we include they will dependably deliver a positive whole number. Crosswise over time and culture, mathematics is by all accounts all-inclusive. It is about the connection to the real world. Mathematics is characterized as the investigation of points, amounts, numbers, structures, spaces and changes.

The exponential function of mathematics could be deceiving, as it starts very small, then goes to infinity doubling, 1, 2, 4, 8, 16, 32, 64, 128, 256. Keep doubling, and without an advanced notice, you would reach a number larger than the number of atoms in the universe. For instance, in the game of chess, the deductive and limited inductive rationale utilized is similar to the valuable arithmetic associated with handling Chess puzzles like the Knight's tour. Every one of the 16-pawns has at most 6 moves before being advanced. It is workable for each pawn to make 6 moves, and at that point, there are presently 16 crosses 6, makes 96 conceivable moves by pawns. There are additionally 30 pieces that can be taken amid a diversion, so the most extreme number of moves satisfying the states of the standard is 126. Expecting that these moves occur as rarely as is conceivable, there are at most 126 crosses, 50 + 50 = 6350, possible moves in a round of Chess, therefore the king is the mathematician in the chess game, who must always stay ahead of the game with the best moves. In math equations, every time you add a new variable, the complexity increases exponentially; therefore, digital transformation becomes all the more complex and very challenging.

As the risks introduced are multi-fold, there will be major consistent changes in our reality that are rapid in numerous industries of the world through digitization. Change brings dangers and openings. A change can mean a hazard for an organization if the market changes and there is never again a requirement for its products.

Change additionally brings numerous chances and potential outcomes for new developments. Today what is to come is being formed rather than the future doing the molding. Therefore, well-established data on conceivable patterns, future improvements, and megatrends will in a few years affect all territories of society and the economy around the world. Since trends are multidimensional, a high level of consideration is paid to them. Multidimensional implies that numerous patterns that one does not find in one's very own industry at first sight are all things considered regularly applicable because of chain and connection impacts.

## WHEAT AND CHESSBOARD PROBLEM

If the chessboard were to have grains of wheat put upon each square with the end goal that one grain was put on the primary square, two on the second, four on the third, etc. (multiplying the number of grains on each consequent square), what number of grains of wheat would be on the chessboard towards the completion? The issue might be tackled utilizing basic expansion. With 8x8 blocks, or 64 squares, on the chessboard, as the quantity of grains duplicates on progressive fashion,

1, 2, 4, 8, ... etc. for the 64 squares. The overall number of grains rises to 18,446,744,073,709,551,615, beyond any prediction or anticipation. The deception here, the function of duplication starts very small, 1, 2, 4, but then grows very quickly beyond imagination. This activity can be utilized to exhibit how rapidly exponential groupings develop, just as to present examples, zero power, capital-sigma documentation, and geometric arrangement. Refreshed for present-day times utilizing pennies and the theoretical inquiry, "Would you rather have a million dollars or the whole of a penny multiplied each day for a month?" the equation has been utilized to clarify intensified premium. (For this situation, the overall estimation of the subsequent pennies would outperform two million dollars in February or ten million dollars in different months.)

## Navigating the Digital Transformation

The Digital Matrix is a ground-breaking area which clearly states the powers that are driving digital changes all over the world and provoking each business to reevaluate its position and re-imagine itself for the near future. As the fate of each industry is advancing, the future is nearer than we might suspect. The powers driving the change are showing in the present perplexity to give us a completely changed business world in a couple of brief years. In 2007 there was just 1 innovation organization in the Fortune 500 top 10. Today the main 5 are advanced innovation organizations and one year from now, this is probably going to be 9 of the best 10. The remarkable development and impact of these technological giants is changing the business scene for us all. By 2025 each business of scale in any industry should have a grasp on technology and its application in business ecosystems. The business goals of advanced change are experimentation, impact and innovation.

Digital transformation has become very necessary for all sectors of the economy. It is an unpredictable, multifaceted procedure that speaks to the working environment and changes the influence between all aspects of life. Digital transformation has changed almost all aspects of human existence: how we shop, tune in to music, read the news and book appointments. Life is gradually becoming easier for people in the digital era as compared to those who lived in the early revolution of human development.

## The Matrix of Digital Transformation

For as long as a couple of years now, it seems like you can barely pivot in a specialized meeting without hearing individuals talk about digital change. The most significant piece of advanced change is really the general population included. Utilizing innovation to build effectiveness by changing paper-based procedures to electronic ones, which composes paper archives, streamlines business procedures, and enables organizations to set aside extra cash and become increasingly profitable. Sorting out paper records and streamlining business forms offers an extraordinary guarantee in sparing organizations cash. Organizations need to tap the maximum capacity that innovation holds. Digitization has made it possible to convert paper into an electronic file organizer for computerized filing and recovery, categorize reports and oversee data in a protected manner. Digital transformation is about innovation; however, the change is about individuals and persuading societies to change their outlook towards technological advancement. Technological change is similarly as significant as the innovations, procedures, and foundations included. As the strain to improve learning and technological skills keeps on increasing, professionals in every industry need to grow better approaches to integrate technological changes into solving world problems.

The ultimate goal of Digital Transformation is to help organizations remain competitive, and improve their services both online and offline by offering increased value to their customers, partners and stakeholders. The pressure on organizations to develop, and in an ever-changing and unpredictable environment, has made digital change a top need for organizations in every phase of their operations, also huge investments are being made to keep up with the pace of technological innovations. Digitization requires building up an expansive cluster of innovation-related resources and business abilities, that can help move an organization along the way toward turning into an advanced high technology enterprise.

## Digital Transformation in People

A major key to digital transformation is people. The social side of the inquiry is simply imperative to digital transformation. People are simple animals who have day to day varieties in their sentiments and practices; they frequently have diversions and twists in what is heard and how things should be; therefore, making them the best source to gather needed information before any digital transformation could ever emerge. Effective advanced change is actually about individuals. Before any organization can digitize, there must be need to transform the people in that organization. Change is hard, people frequently need to experience an evaluation before they focus on embracing themselves with the needed change of digital transformation, and they have to know how the progressions will influence them. People comprehend a lot of information, more so why they must be consulted before developing any technological innovation in any industry. Digital changes require consistency, as engineering and business forms need time and specifics. Before managing the innovative component of digitization, the human dimension of this procedure should be painstakingly overseen. Change for individuals happens at each dimension of their interaction and association with others in their environment. Individuals always analyze what they would do in any given situation; in what manner will they carry on, work with others, create, and develop themselves as well as other people in both the digital and physical world. These are more reasons that a portion of the human factors should be integrated into the way of life of digital transformation. For instance, trust, humility, collaboration, teams, self-learning, transparency risk, and listening to the technological needs should be fully put into consideration before upgrading technological innovations in order to enjoy productivity from technological advancements.

## Digital Transformation in Process

In view of present and future movements, in this digital era, there are not many things more critical to an organization's procedure than making a powerful advanced change process. When planning their IT infrastructures and methodologies, officials, tech pioneers and change supervisors must record for both the innovative and human segments of progress by evaluating the present technological needs.

Administrators must clearly characterize what is going on inside the market that requires their organizations to be fully ready to embrace digital transformation. Digital transformation is drastically changing the worldwide business scene as all the business verticals are utilizing the key switches of Digital Transformation including Social Platforms, Internet of Things, Mobility, Cloud, and Analytics to build the execution and reach of their business. The reception and utilization of these problematic advances, in the perfect time, and at the ideal spot, empower associations to expand business readiness, give better client experience and distinguish new business openings. Digital trends should be added to organizations' objectives with significant research on the best possible time to fully upgrade technologies peculiar to business operations. As much as new innovation enables organizations to remain productive and aggressive, the best method to approach the advanced change process is to consider new innovation as a device to accomplish a more noteworthy key objective. Embracing innovation is not just about including something sparkling and new but recognizing the right innovation. Technological changes must occur where they are highly needed and would bring about great positive impact with the most minimal hazard. Digital transformation in organizations must be will thought, and properly planned for, with capacity and sustainability. The process and impact of digitization, digitization require comprehensive strategies, and predictive anslysis of possible outcomes; it simply needs to work with the organization's specific needs and ensures that those objectives are fully met.

## Digital Transformation in Technology

In the present market, it is the advanced change slants of high technology firms that is molding digital transformation, as most advanced innovations give conceivable outcomes to productivity additions and client closeness. A Digital Transformation process incorporates audits and updates to Goals, Objectives, Strategies, Processes, Metrics, and Technologies. The reason for this change is to get the most recent innovation that is currently moderate and accessible, now more than ever, that would benefit the organization on short, medium and long term periods. The accompanying rundown of top innovations are being executed today in different organizations to carefully change the manner in which their representatives work and significantly improve for improved customer experience. Organizations are advancing towards information-driven basic leadership, by making the procedures and capacities to process information dependably continuous. High tech organizations are working tirelessly to make the idea of information-driven choices a reliable alternative in advanced change through the improvement of the framework, programming, and administrations that can genuinely convey the guarantee of information. Clear necessities and desires combined with a test to tackle a genuine issue gives the earth its genuine advancement to prosper to realization. Innovation is manufactured one practical piece at a time; with definite parameters, it encourages clear correspondence and language around the innate dangers of advancement. Emphasis is major to advancement. Digital transformation includes utilizing advanced technological solutions to revamp a procedure to become progressively productive or compelling. The thought is to utilize innovation not simply to repeat a current technological infrastructure but to utilize innovation to change that structure into something essentially better.

There are information issues that organizations need to address throughout the following couple of years. One is that most inheritance frameworks are singular storehouses of data and were never intended to cooperate. The volume of information will exponentially develop when you overhaul from old heritage applications to new present-day innovation frameworks. Business choices are drastically accelerating and you have to give your group the key detailing instruments that give moment data to settle on the best choices for the organization. Cloud and Mobile Apps are required nowadays for most experts to be gainful. Organizations need to be updated with proactive and receptive measures needed to ensure digital security at all times, even as they embrace the latest technological advancements.

Learning to navigate the new complex world has proven to be so vital and extensive. In the business world there are strategies to the digital transformation, and for the purpose of this book, we will be discussing the strategy of transformation outside in and the transformation roadmap inside out. In the next aspects we will be discussing the strategies of digital transformation from a business perspective. The next chapter highlights the strategy of transformation outside-in.

# CHAPTER 6

## THE TRANSFORMATION STRATEGY

In the outside-in perspective, organizations place major emphasis on the needs of customers instead of the product with continuous expansive thinking. They are completely immersed in the thought pattern of clients, getting their feedback and developing the right product for the them based on their needs. Inside-Out is characterized by activities by organizations that are inside driven. For example, user experience improvements, multi-channel access, and frameworks move up to upgrade straight-through preparing. Digital Outside-In activities are remotely determined with the objective of expanding organizations' advanced capacities. This process depends on the core abilities of the organization to drive change, product advancement and business development instead of outer impacts, for example, market, rivalry and client inclinations. An organization accomplishes more prominent efficiencies and adjusts all the more rapidly to evolving conditions. As organizations set their focus on the future they appear to be categorized as those guided by inside-out reasoning and those by outside-in considerations. The job of the organization guided by inside-out reasoning is to accomplish sustainable and gainful development. Outside-In reasoning spots the organization's objective reasoning at the core of its plan of action, with the view that commitment towards a higher objective will create positive business results also. Without interest in development, testing, and making a stride outside the corporate system and into the brains and hearts of clients, a business would not be prepared for the minute when markets open and their chance to beat rivalry and expanding the piece of the pie and brand devotion arrives.

For instance, Ford motors Alan Mulally drove Ford Motor Company through major extreme change. He was guided greatly by the concept of inside-out reasoning which made Ford do whatever it would take to develop their capacities. He cut down expenses to the bare minimum, radically diminished the organization's work power and put a premium on advancement and product quality. His goal was based on total quality improvement for their company as a brand. Ford, at some point in time, asked people what they wanted and the reply was a faster horse. It begins by asking what an organization can do with existing assets, and hopes to streamline activities through right estimating and minimized spending. Another case of outside-in strategy in business is Toyota. The organization moved far from gathering client needs toward the inner objective of beating General Motors and expanding development. Outside-in reasoning diverts organizations from their actual motivation behind driving client esteem to understanding the contrast between what you make and what individuals need. Outside-in intuition implies that organizations take a look at their business from the client's point of view and along these lines configure procedures, devices, and products, and then settle on choices dependent on what is best fit for customers.

Linear vs Exponential functions — the difference:

In basic mathematics, linear and exponential functions are pivotal and foundational aspects that are later on built upon to achieve greater results even in more complex situations or problems.

Linear functions simply are those that progress or change at regulated intervals and constant value added or deducted. The linear equations in algebraic problems follow a simple pattern which when the first sequence is understood can be used to solve the others using the same noticeable arrangements.

Exponential functions, on the other hand, are the ones whose value is always proportional to the value of the function. Exponential functions do not carry base powers alone but also squares and other powers.

Change in linear functions is observable at regular intervals while change in exponential functions changes by common ratio over a common interval.

The sustainability of industries and their productivity is dependent on these two. While linear functions might be responsible for the in-house factors that affect production, exponential functions might be responsible for the external factors that affect production. In the Constable unit who could observe that the in-house factors that affect production could be: *people* + *cost* + *technology.*

These might have easily understandable and observable behaviours and can be capitalised upon to maintain a balanced policy for industrial process. Internal factors include workers' wages and cost of keeping up available technological tools.

External factors which are exponential in nature, are with the formula: ***Customers + industry + competition***

The unpredictable obvious factors make it hard to formulate it definite charts in order to utilise the information for industrial precaution and policy-making. In essence, the policy of the competition and taste of the consumer which has no handy outlook, might affect the production — this is because the production is an exponential function to the external factors. And anytime new variables are added, the complexity gets increased.

To maintain a great balance of linear and exponential functions and production, the companies could employ an outside-in strategy which will include:

Industry and trend — The companies must be well acquainted with the waves of the moment in the particular industries and utilise that knowledge in defining the destiny of their external factors by themselves. That way, they are not on the defensive against external factors, rather, they have launched an offensive to their favour, which will thereafter become an external factor for the competition to manage.

Customer based production — The taste of the consumer should dictate the designs and service rendering of the companies. This allows the company more ground in the industrial scale as well as allows for the direct communication with the customers as to improve on specific products.

Competition & Partnership — In the face of tough competition, the companies might be forced to do either of three things: to upgrade their production to become better, to strike a partnership with another company to face off the big bully or to quit the business. Intense competition should be turned into immense opportunities.

## THE INSIDE-OUT STRATEGY

This is a strategy that brings the companies involved into self-questioning and self-betterment. Some of the strategies that can help sustain a company as well as limit the telling factors of exponential functions include:

**Business process:** — The business process of a company that would function well in a perfect market would most necessarily be thorough. The processes must be clear and such that can be easily crosschecked when needed. It must match the technological as well as legal requirements — no axe evasion, no inferior machines. This business process should also be concealed from the competition in order to avoid shooting oneself in the leg. As much as possible, a good company should hide its grand plan from any sort of competition.

**Insights and consumer's values**: — The backdrop of a company's production must be painted with the values of the consumer, especially in a traditional society. However, as values are relative, the insights of customers on the modifications and presentation should be adopted to affect an all-round success in production.

**Employee's Experience**: — The experience of the employees must also count as an additional benefit to the company in beating the external factors that contribute to exponential mishaps. The company needs more experience than enthusiasm. A company that would survive every season regardless of external factors needs more hands to weather the storm and experience counts in industrial process.

**Technology and data**: — This is another inside-out approach to managing complexities. The company's database must be well ordered. This helps to observe and correct glitches and also make certain needed permutations. Consider also that the competitions might not be traditional, they might also be internet competition. The internet could also affect the industry and as such, the companies must strike a balance between online and offline service rendering and production.

The human factors affecting production and distribution must be augmented to becoming more efficient in keeping up with the demanding and revolutionary world around. The telling effect on the company would be maximum productivity and maximum satisfaction. Skill acquisition should be a continuous process and the people involved must be ready to embrace new technological changes as well as technical manoeuvre. The industrial process should also be refined beyond the traditional space in order to connect with a ready customer base and also maintain relevance in the e-world. The companies should also adapt ready-made solutions to suit the pressing needs around them.

## BUSINESS OUTSIDE-IN STRATEGY

Most organizations are consumed with their own internal affairs, and do not pay attention to what is happening in the industry or around them. That is when they miss the market transition due to the forces of change. Imagine surfing in the ocean, without the proper visibility and understanding of the ocean and water. From an Outside-In strategy, investor confidence comes from tuning in and offering some incentive to clients and helping them show signs of improvement while giving a consistent client experience. When clients are not happy, investor confidence would decrease the business. Outside-in strategy helps organizations understand clients' needs better and gives them the best product that would serve their needs. Businesses' outside-in believing strategy serves as a link between social customer relationship management (customers, partners, and stakeholders) and design thinking in developing business capabilities. Developments are effectively fused into collaboration designs among organizations and clients. Innovation improvements keep on improving strategies. For executing business, easy trade is on the rise ascent. Buyers need to utilize digital transformation like a universal remote gadget that makes life simpler. The outside-in approach is a deliberate and creative approach to evaluate and upgrade any organization. It guarantees strategies to remain in front of the market and be receptive to changes. An outside-in methodology enables organizations to protect against various dangers, by staying above market and environmental changes. Understanding clients from the outside-in context ought to be viewed as a core objective for any organization. The bits of knowledge, particularly customer's online behaviour and perception, are enormously profitable. Organizations that consider the outside-in of their customers are bound to adjust and flourish. The more accurate the view, the better the strategy will be for organizations.

Having highlighted all that is involved in the strategy of transformation outside-in, it is appropriate that we consider the digital transformation roadmap inside-out. This will help us get conversant with how people and processes are changing with the magnitude of digital disruption as highlighted in chapter two above. So, in the next chapter we will be breaking down the transformation roadmap inside-out.

# CHAPTER 7

## THE TRANSFORMATION ROADMAP

People are prone to have suspicions towards technology since it undermines a significant number of convictions in their conditions. Understanding the human component of progress is critical to accomplishing digital transformation effectively. It is fundamental to make people who are longing for change the chance to mold their interest and capabilities in the areas they want them. Thereby, advancing the procedure of progress needed in digital transformation becomes successful. The advanced technological change applies new innovation to improve all parts of the client experience and expectations. This incorporates creating or embracing the correct innovations to modernize products and procedures needed to effect digital change. This goes past structure of technical abilities; it requires a profound comprehension of what present-day customers need and reevaluates digital strategy to meet the need. Innovation is driving profound changes in organizations strategy, workforce productivity, and improved customer experience. Smaller organizations are growing new ranges of technical abilities to comprehend the potential for innovation supported advancement by changing operations, key partners, stakeholders and collaborations within both internal and external business environments, in order to make innovation a center competency.

Mobility is the eventual fate of everything with the speed and adoption of the Internet of Things and smartphones. Our lives and inclinations have obviously changed; everything is gradually getting connected from the physical world into the virtual world. Innovation is utilized to take care of conventional business issues and generally change the way the business works. The most ideal approach to draw your digital transformation motivation is to gain from digital giants like Amazon, who is the current market leader in the retail sector. Ten years back nothing proposed a danger to physical stores but today most stores want to go online, advertise and keep customers updated with their latest stock, either through their website, mobile application or newsletters. In 2005 Amazon was only an online book shop with humble deals. Today it is worth more than all real physical stores in the United States combined together;. While the majority of the recorded retail locations saw a misfortune in sales, the business for Amazon was just blossoming where it had experienced an increase of

1,934 % in sales and went directly to a standout amongst the most profitable organizations on the planet. Most physical stores decades back failed to see all the noticeable signs of the digital transformation roadmap. On their way, they got into the digital business model very late as compared to Amazon. Creating transformative applications at speed requires a solid hierarchical culture and close joint effort among everybody engaged with the task. Advanced development occurs through collaboration. As organizations keep on embracing technology and developments, the higher their capacities become to perform and innovate. Joining speed with the capacity to scale up and alter the applications as and when required can guarantee set objectives be achieved promptly.

**Changing People, Process, And Technology**

We frequently hear the management consultants reference the expression, people, process and technology as a method for clarifying the basic achievement factors for hierarchical change in a time of automation, AI, and machine learning. The expression started from Harold Leavitt's 1964 paper "Applied Organization Change in

Industry". In it, he places a four-section "Diamond" model for making a change in an organization. Leavitt in his paper said that structure represents how people in an organization are organized, while tasks were the duties assigned to each worker in an organization. People represented the workforce and technology represented the tool used by organization workforce to carry out their individual tasks. Since the distribution of that paper, organizations have been able to integrate structure and tasks down to Process, to what workers do. The People, Process, and Technology model is ageless as a result of its simplicity, yet one of its eccentricities is that it reveals to you just what the elements are, not what they do or how they interact.

## How People, Process, and Technology Interact

How do these entities, seeming discrete in Leavitt's model, work with one another, and how would we utilize them? People have to work so that they can enjoy their life. How they do their work and what they do their work with is the key inquiry. Even in the time of AI, people are still responsible for the outcome of machines even in this digital era.

The process enables people to improve the quality of their work and individual capacities. Process characterizes and institutionalizes work, thereby keeping people from reexamining the wheel each time they start working.

Technology enables people to do quicker, increasingly creative work, particularly in the time of artificial intelligence. We hand off repetition and mechanical undertakings to machines, from coffee to transcribing of speech so as to save our time for more imaginative, psychological undertakings.

# DIGITAL ROADMAP FOR BUSINESS

Technological innovations will not deliver the desired benefits unless they are used in the right way, in the right place and by people who know how to use them. The key here is to get the planning right from the onset of digital transformation. The key to deriving long-lasting benefits from innovation is how digital technology performs at the point of use in relevance to the business, and the speed of execution. Whatever technology invested in, whether for the purpose of connectivity, logistics, navigation, tracking or energy saving, the goal is to derive maximum benefit from its usage. This could be quantified in several ways, such as cost savings, time efficiencies, reliability, reduced emissions, customer service, safety and more that makes a particular innovation better than the previous model. A digital roadmap serves as an outline for needed actions that makes it possible to align digital initiatives with business objectives. A digital roadmap clearly states the difference between various plans by an organization and a cohesive strategy that makes business goals a reality with stipulated timelines and reviews. Visual representation of digital business roadmap with action plans highlighted is advised due to the fact that it would help the organization know the path it needs to follow to get to the desired result. A brief explanation of plans split by milestone stages with an overview of the main aims of each stage is needed in visual forms to keep the digital roadmap plan alive and running. The digital business roadmap should be a concise important document that is easily accessible to anyone within an organization.

Technology keeps evolving with the constant transition into cutting edge solutions for businesses. It is not the change that keeps some traditional companies out of sales but their failure to transition into a digital business. Today most business owners have social media platforms that help them connect to their customers and also get feedback through their social media platforms due to the fact that the business world never stops evolving. Digital roadmap serves as a global positioning system (GPS) that shows a company's digital strategy in line with business goals and objectives. Digital business roadmap should be a live document that is flexible and can develop with the business, giving a steady reference point to keep the digital procedure on track and lined up with the business objectives. Even as those objectives advance after some time, it should be looked into regularly and refreshed in accordance with constant investigation, changing business necessities and impact from external environment. The digital roadmap should be established on clearly explained objectives from the beginning and should incorporate key execution markers to quantify the advancement of the digital system against those objectives. It should work in eager yet feasible achievements and checkpoints for surveying the progress made, showing the method of reasoning of the proposed advanced procedure. A valuable initial step is to utilize strengths, weaknesses, opportunities, and threats (SWOT) investigation to distinguish where the holes are in business procedures, operations, execution, and capacities.

## PEOPLE TRANSFORMATION

People are fundamental in the achievement of digital transformation. Every change procedure must begin with people in the society. In business, it must begin with employees. Just as digital transformation is important, the people part is greatly as important as the technology, process, and infrastructures involved in learning and development. Digital transformation starts with people due to the fact that there has to be a change in mindset of people firstly before a technological change could be accepted. For instance, before the advent of the World Wide Web, there was a massive acceptance of personal computers worldwide which gave rise to the need of people wanting to connect globally through the internet and social media platforms. People come first, before any technology transformation which is why they need to be aware of technological shifts and advancements. They need to know the benefits and challenges any technological upgrade would bring. For instance, the European government is concerned with the privacy law of its citizens. The issue of privacy is one of the disadvantages of social media platforms, although it is an asset to some marketing companies. It could sometimes make people uncomfortable due to the fact that their privacy is being invaded and most of their activities are being monitored by high tech companies. Digital change is rapidly gaining access into different industry segments, as organizations endeavor to utilize innovative headways to remain aggressive and relevant in an ever-evolving world. However, numerous organizations taking a stab at digital change neglect to understand that the change additionally requires an interest in the general population utilizing it.

Digital Change should never begin with innovation, as organizations hope to transform. It should all be centered on the human hearts and brains of people in the society. Their hearts determine acceptance while their brains figure out the benefits of technological advancements such as mobility and disadvantages. A social move requires tolerance and ingenuity just as a community-oriented effort from all gatherings included. Change must begin with the involvement of people in the ecosystem. People must be ready before any advanced tech is brought into spot. The most critical errors most technological enthusiasts and inventors make is to utilize new innovations as an endeavor to suppress human action, as opposed to developing a tool that would enhance human capacity and productivity. People are naturally made to have control over their environment. Technology can only help them control their environment seamlessly.

# WORKFORCE TRANSFORMATION

Innovation does not drive change; it is the manner by which that innovation is being embraced and utilized by the general population that characterizes digital transformation in any industry. Workforce change is basically a hierarchical culture improvement activity which requires comprehension and support in activities that would bring about quality improvements at every given phase of organizational goals and objectives. Organizations need to consistently change the workplace environment to be open for creativity and innovation that would boost workforce productivity level. It is the staff that will decide if an organization is effective or not, and workforce change can hold tremendous advantages for company directors, executives, stakeholders and the company in general. The world has changed tremendously; the exponential development of profitability, data sharing, versatility and cooperation is reshaping the business world quicker than ever before. The objective of workforce transformation is to determine the most incentive for minimal measure of time and resources. Workforce Transformation would give rise to organizational future development in a quickly changing business condition by enabling more done when human resources are more coordinated and highly trained in tech-related skills. The more the people are skilled in technological areas, the more productive and efficient they become and the more organizational profitability would grow. Where data lives and how it is overseen is fundamental to training workforce not just on how to use modern technologies but also on how to protect themselves, and the technology is very critical in the safety of the organization. Workforce transformation must come with the goal of having workers protected from potential risks associated with modern technology by going through manuals. For instance, most high-tech machines in production lines come with the hazards of radiation staff who would function in such departments. This needs to be fully insured and informed of the hazards technological devices can be associated with. Workforce transformation should be integrated with training, safety and security before workforce productivity would be felt within an organization.

## TECHNOLOGY TRANSFORMATION

Technology can make the work simpler, however it cannot make people reduce errors or make them become highly proficient in every area of their lives just as technology cannot make colleagues and friends communicate with each other if they do not enable their device to communication. It requires effort, learning and control to achieve the desired result. Technology requires institutionalized and systematized forms that are obviously settled and surely understood. No measure of innovation can compensate for the weaknesses of an understaffed division in any organization regardless of how much certain bits of innovation can streamline procedures and make them simpler for staff. When people, process and technology are correctly put in place, innovation circle and creating the ideal outcomes would gradually become successful in a digital transformation roadmap.

Innovative technology applications are the different ways innovation can be utilized in making financially valuable products. The use of innovation apparatuses and gadgets in the instructing and learning forms include the utilization, learning, aptitude and skill in the utilization innovation in tackling issues or performing explicit capacity. For instance, an application designer or computer programmer can compose their code as a local, web or half breed application and these terms can likewise depict work area applications. The engineer codes a local application to keep running on certain hardware equipment, the Photos application on Mac OS X is written in Objective-C the same language that Mac OS X employs. People can get to a web application by means of an internet browser such as Google Chrome, Mozilla Firefox, Internet Explorer and several others. Likewise, a computer programmer can code in several programming languages, including JavaScript, CSS, HTML and several others. With the technological transformation in all sectors, the possibilities to human creativity are endless.

The fields of science and technology are commonly evolving as new innovations grow the scope of science and engineering, permitting the investigation of domains, already distant, to become reality. The transformation in technology as discussed in the chapter following the inside-out and outside-in strategies can further be analyzed from the Information Technology (IT) and Operational Technology (OT) perspective. The convergence of IT and OT will help broaden our understanding of the horizontal scope and magnitude in the transformations and disruptions in technology processes as related to businesses and organizations.

# CHAPTER 8

## THE CONVERGENCE (IT + OT)

As discussed in the previous chapter, Digital Transformation will cover the transformation across Business Processes, People, and Technology. It will be very important to address the digitization horizontally across the entire organization or company. Digitization will transform horizontally across every function in the business, as well as the very way employees conduct their functions. This chapter will focus on providing detail around bridging the divide between Information Technologies and Operation Technologies.

## OVERVIEW OF INFORMATION TECHNOLOGY

Information Technology (IT) is the utilization of computers and programming software to control information or data. It is the discipline in charge of data storage, data protection, data processing and transmission of data when required. It is also referenced as ICT, Information Technology (IT), or Information and Communications Technology (ICT). IT covers any type of innovation, that is hardware or system utilized by an organization, institution, people, foundation, or any association which handles data. The term Information Technology was instituted in the late 1970s to explain the link between computer technologies and information handling. The term information technology was first introduced by Thomas L. Whisler and Harold J. Leavitt in a 1958 article distributed in the Harvard Business Review. Decades back, 50 years to be precise, only few individuals who worked in large organizations like banks and medical clinics had access to stored information around them, while computer programming and software development were taken care of by mathematicians, computer scientists and researchers since they were too complex to even consider being managed by different individuals and experts. At the time, there was no such thing as a degree or courses in computer science or computer engineering, organizations had no requirement for their IT divisions. There were no interconnected systems, servers, or complex PC frameworks. Telephone, fax, and standard mail were the fundamental correspondence channels. As the IT business developed from the mid-twentieth century, computer operating capacity progressed while gadget costs and sizes of devices fell lower. As time passed, innovation extraordinarily progressed, and pretty much everybody in the western world realized how to utilize a Personal Computer (PC) and saw the need to have one. Likewise, organizations began seeing the advantages of data frameworks. The world rapidly moved into the data age.

Information technology has been around for a long time. However, it is imperative to find out about how we came to the heart of the matter we are at with innovation today. The pre-mechanical age is the most punctual period of data innovation. It tends to be characterized as the time between 3000B.C. and 1450A.D. As letters became more popular and more individuals were recording data, pens and paper started to be created. It started off as just stamps in wet earth, later paper was made out of papyrus plant. The most well-known sort of paper made was presumably by the Chinese who made paper from cloths. Since individuals were recording a ton of data, they required approaches to keep it all in changeless capacity. This is what led to the creation of books and libraries. The mechanical age is the point at which we first began to see associations between our present innovation and its predecessors. The mechanical age can be characterized as the time somewhere between 1450 and 1840. A ton of new inventions were created in this period as there was a vast blast in enthusiasm with this territory. Inventions like the simple PC were developed. Blaise Pascal created the Pascaline which was a mainstream mechanical PC. Charles Babbage built up the difference engine which organized polynomial equations utilizing the strategy for finite differences. The electromechanical age can be characterized as the time somewhere between 1840 and 1940. This was the beginning of media transmission. The phone, a standout amongst the most well-known types of the inventions, was made by Alexander Graham Bell in 1876, and the radio was created by Guglielmo Marconi in 1894. These were amazingly critical developing inventions that prompted huge technological advancement in the field of information technology. The first digital computer was created by Harvard University in 1940. This PC was 8ft high, 50ft long, 2ft wide, and gauged 5 tons – it was huge. It was customized utilizing punch cards. It was from immense machines like this that inventors started to research how to make smaller computers usable by organizations and eventually in every home. The electronic age is where we are at present; it tends to be characterized as the time between 1940 and the present time 2019.

With the move to the paperless office, the advent of computers and smartphones has increased with a high number of internet usage on a daily basis. IT has turned into an easily recognized name and discipline that so many young people want to be part of or want to start a successful career in. Nowadays, by far most organizations would not get by without the utilization of data frameworks. They are sending messages through email, text or through their social media platform, advertising through digital marketing, managing their website and online presence. From the enormous aggregates to the littlest self-start ventures, there is no uncertainty that organizations everywhere throughout the world depend intensely on data frameworks in the relentless and aggressive universe of business. Those with the best and most current innovations are the ones that flourish and the ones that are stuck in the past are abandoned. Business applications incorporate databases like SQL Server, value-based frameworks, email servers and web servers; these applications execute modified guidelines to control, solidify, and scatter or generally influence information for business reasons. PC servers run business applications. Servers communicate with customer clients and different servers crosswise over at least one business system. Capacity is any sort of innovation that holds data as information. Data can take any shape including document information, mixed media, communication information, web information and information from sensors or future plans. Capacity incorporates Random Access Memory (RAM), a type of computer memory that can be accessed randomly. That is, any byte of memory can be accessed without touching the preceding bytes. RAM is found in servers, PCs, tablets, smartphones and other devices, such as printers. IT models have developed to incorporate virtualization and distributed computing, where physical assets are disconnected and pooled in various setups to meet application necessities.

## OVERVIEW OF OPERATIONAL TECHNOLOGY

Operational Technology (OT) is the integration of information technology which includes hardware and software to control physical processes, devices, and infrastructure. Operational Technology is a term used in many different ways by many experts in the industry.

Some refer to it as being the information and communications technologies, systems, protocols, and infrastructures. While others refer to it as all kinds of technologies and machinery that are being used in any business to run the business. For example, in a hospital, the magnetic resonance imaging (MRI) machine would be an operational technology, and a factory machine in manufacturing would be an operational technology as well. Operational technology is basic in Industrial Control Systems. It is utilized to control stations or open transportation terminals. As this innovation propels and unites with organized tech, the requirement for OT security develops exponentially. Operational innovation is utilized in Public Infrastructure as a framework that enables administrators to control the progression of water through a reservoir conduit or channel. In the transportation segment, OT is utilized for remotely checking and controlling an icebox vehicle while it is on a train. In urban communities, activity innovation enables a city to deal with an arrangement of chargers for electric autos. OT enables solar system boards to be remotely upgraded including functions like cleaning; in an office OT permits buildings to have optimized robotized controls for warming, ventilation and cooling.

According to the Harvard Business Review, the field of operations has experienced some significant developments over its history. The industrial revolution of 1800 gave rise to the field of operations technology. The field took off as the advanced economy rose up out of the new wonder of the capacity of manufacturing. Trailblazers like Eli Whitney of the cotton gin drove the route with the promotion of assembling frameworks that changed a craftsman economy dependent on filling and fitting parts. The ideas of exchangeable parts empowered another type of industrialist to create and sharpen a particular modular system for production in which singular segments could be made autonomously and at a larger scale. This step by step prompted the ideas of coordination, supply chains, and sequential construction systems, and shaped the establishments of the American System of Manufacturing, which developed in the first half of the twentieth century and was popular in the 1950 and 1960 era. The field of operations research exploded in 1960 due to higher demands to break down and enhance the progression of goods and information in the manufacturing sector. The utilization of these techniques soon spread fast beyond manufacturing assembling into other sectors like banks to electric utilities. It also prompted the foundation of service management and service operations as center subjects in the field of operations. The advancement continued in 1980 and 1990, as new ages of digital innovation started to change the basics of working and productivity. This led to the field of management of organizations conveying software and programming-based products like Microsoft and Yahoo Corporation requiring operations.

Digital technology has empowered activities like the management of information which has greatly increased work efficiency. From the time of the primary business, IBM centralized computers in the late 1950s, PCs have driven expanding productivity into assembling and administration of organizations over time. The ongoing universality of digital innovation and its scope of utilization in web administrations, mobile, and now the internet of things implies that the advancement and delivery of software services is beginning to change the very texture of our business and working conditions. Digital technology is changing the nature by which control is characterized and delivered. The design, management, and deployment of software has become integral to an association's working model. Digital innovation is additionally empowering totally new working models that are progressively open, appropriated, and shared globally, as these new models have empowered nearly 9 million independent software developers to contribute mobile applications to the iOS and Android versatile mobile platforms. They've empowered Uber's 2,000 interior workers to deal with the unpredictable coordination of 200,000 drivers. Furthermore, they have empowered WhatsApp to develop to more than 450,000 clients with less than 30 people in staff. The structure of tools for development, working framework APIs, or the client onboarding process for a mobile application have moved toward becoming urgent and operating excellence, production planning, and inventory control.

## THE CONVERGENCE OF IT + OT

Communications, security, building controls, mechanical procedure control, assembling, manufacturing and innumerable different fields started before there were economical microchips and fast communication protocols. Computer and communications figurations have always been related to controlling physical devices as their essential core interest. Specific computer processing and communications were frequently required to meet unending needs that were altogether different from the necessities of business computing tended to by IT. Present day control frameworks require critical ability in embedded frameworks and security to design, structure, fabricate and maintain system engineering methods and formal lifecycle development processes. Industrial applications require a multidisciplinary approach involving subject expertise, software, computing and communications. Operational adequacy requires coordinated business planning, portfolio management, project delivery, and continuous tasks administration. Of these difficulties, a standout amongst the most basic is security. Security and digital assaults with IT/OT can prompt annihilating results, which may incorporate power failure, power outages and loss of secret data and records with constant endeavors to devise policies, techniques, and organizational culture when these difficulties can be survived and avoided totally. At the point when people consider the innovation inside their organization, they ponder information technology (IT). Actually, venture innovation, likewise, incorporates operational technology (OT), particularly in modern firms. OT is frequently and considerably more mission-basic than IT as OT fuses all the physical framework that underlies numerous organizations, including power plants, generation lines, vehicles, trains and flying machine motors, HVAC gear, oil and gas rig hardware and numerous others. This framework is comprised of complex hardware and is progressively being associated with systems. Operational technology is pertinent for most organizations; however, particularly those in modern segments like assembling, oil and gas, utilities, administration, and repairs.

The convergence of IT and OT frameworks is a characteristic wonder for organizations setting out on the advanced change venture. Be that as it may, as most technological advancements do not come without hazards, this hazard is particularly common in ventures like assembling and manufacturing with worldwide inventory network impressions and little edges of errors. Information technology (IT) and operations technology (OT) exist together in numerous offices; however, they frequently do not cooperate. It is indispensable to forget about the significance of combining IT and OT to upgrade activities and to accomplish the advantages of their combined usage. IT/OT coordination results in a more streamlined and less expensive innovation organization. In conveying IT procedures and capacities to OT, IT needs to perceive the one of a kind prerequisites of basic control frameworks and the equal procedure abilities that OT accommodates in the designing and activity of basic control frameworks. Operations technology is the utilization of PCs, programming, equipment and other media transmission gadgets to perform business activities. OT is related to front-end, field-based gadgets used to perform genuine activities. These gadgets are chiefly situated in office areas and server rooms. IT is related with back-end capacities used to perform different business activities; IT and OT convergence implies incorporating operational innovations, for example, SCADA, remote terminal unit, programmable rationale controllers, and meters and sensors, which work continuously or closely with IT frameworks. IT and OT groups have regularly had the independent, unlimited authority of their techniques and spending plans. They have huge experience and fitness in their own spaces. Intermingling does not mean exchanging IT engineers with plant designers or the other way around, it implies building shared belief between the associations to such an extent that they can depend on each other as masters inside a bigger association whose coordinated effort is basic to progress and creating the bridge between Information and Operational Technologies through use cases that deliver on business outcomes and address the business priorities. That is digital transformation.

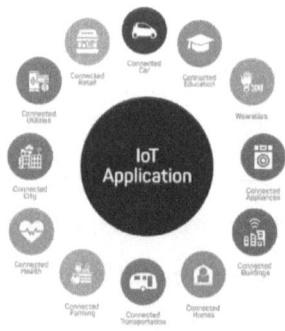

Since the coming of the Industrial Internet of Things (IIoT), specialists and enthusiasts have been discussing the meeting up of IT (information technology) and OT (operational technology). Some consider it coordination, while others consider it a combination. IT/OT combination is the coordination of information technology (IT) frameworks utilized for information-driven computing with operational technology (OT) frameworks used to monitor occasions, processes and gadgets so as to make changes in big business and modern activities. Organizations worldwide will go through $2.1 trillion per year on digital transformation projects by 2021. Truly, IT/OT task departments can work independently, within an industrial manufacturing organization. While OT can keep the plant running easily, IT can oversee business applications from the front office.

## THE FACTORY OF THE FUTURE

The OT/IT intermingling can be seen as a coordination between the venture and implanted systems. OT and IT organizations are awakening to the fact that the best way to advance beyond the game is to partner together to offer coordinated solutions. Traditional factories should get ready for what's to come. In the factory of things to come, value streams are completely digitized. All hardware modules and capacities are represented digitally in real time. Sophisticated software solutions gather, move and transfer process information from manufacturing and logistics to break down and analyze the data collated. This improves all procedures over the value stream, from acquisition to production, directly all the way to the end user or customers. A definitive objective is to increase the level of productivity, profitability, and effectiveness. Data is the most valuable resource in the factory of the future. All systems, including hardware, software, machines and equipment will be well represented by digital data. Profitability, cost/benefit, quality, and productivity will all be estimated and investigated continuously. Only people who deeply understand the digital transformation would be the pioneers in their industry.

## THE PROMISE OF OT-IT

The Internet of Things (IoT) has grown in the previous 5 to 10 years. As indicated by Business Intelligence Statista, there were more than 20 billion associated gadgets toward the finish of 2017. Furthermore, by 2020, the anticipated introduced base will surpass 30 billion, while machine-to-machine interfaces will number more than 3 billion. At this unprecedented dimension of the network, the industry can be scaled significantly bigger unexpectedly, through smarter gadgets. Complex global supply chain network presently depends on digital communications and automation. Information multiplies, from that gathered from the remote terminal unit (RTU) to scanners checking clusters on the production line, to customer data mined by means of social media activities. 80% of supply chain experts believe that in five years' time, digital value systems will be omnipresent. Connecting OT with IT is a door to the next level of business performance. Smarter machines can accomplish more, with supreme exactness. To a more prominent degree, the workforce will now be able to be coordinated to progressively gainful and profitable tasks and work expenses can be cut tremendously.

## IT/OT INTEGRATION

The movement of the technological revolution with the precision that we are experiencing, demands something even bigger — the need to ensure the smooth running of the world as a single heavy machine. The future technological kingdom in construction needs to balance well on both IT and OT in order to strike equilibrium as well as bring about adequate test running for the levers of industries that would survive.

The importance of IT and OT working together is immense. For example, in the engineering industry, productions could be adequately monitored to suit the accurate need of the consumer and also to prevent hassles that come with customized production.

The integration of the information technology and operational technology brings the consumers in direct contact with the producers. Since IT covers the relativity of communication as an informational scope, OT brings along an adequate security buildup. And this brings the two to work together in tandem. As it was, some OT systems were built without consideration of remote access, this therefore, weakened the bond that could come between OT and IT system. However, these could be upgraded to meet the standard of the already fully functional and fast paced IT system.

The management and running of industries are enhanced and they double proofed the mechanical process and brought transparency to industrialization.

IT has the challenges of an outdated system that needs to be synchronized to meet the present industrial internet requirements, while OT also keeps a catalog of information leaks as well as physical safety concerns. While these two possess quite interesting and great prospects and even more possibilities, the convergence of the two is much better for the new internet-based industry.

## BENEFITS OF OT/IT CONVERGENCE

Next-level efficiency: IoT innovations can help join effort inside a workforce, regardless of whether it is training or maintenance activities. Boeing utilized the Augmented Reality (AR) application Google Glass to test efficiency and precision on a complex wiring assembly, revealing time reserve funds of 25%. The task issue rate additionally divided.

**Improved Control**: Wireless communications and mobile devices allow fast information delivery and data sharing. Enabling "work anywhere" workforce, organizations currently can reach instantly their remotest tasks or field workplaces, however unrestricted access and visibility enables constant activities. For instance, service organizations will now be able to utilize dashboards which report live consumption of data, stratified over numerous observing case studies, for example, rural versus industry utilization or use by region. The framework may likewise utilize predictive algorithms to the caution of over-loads, imbalances or blackouts and automate the expected changes to the system.

**Deeper Insights**: Integration opens a window to display upgrades in defined regions of the business. Utilizing explanatory devices, aggregated just as live information can be demonstrated for development. OT-IT cooperative energies can create upgraded generation streams, seed customer collaboration opportunities or give researched information on consumer data very fast with accuracy. The keys are to organize applications to understand how to reduce information towards issues that give business-changing insights. Knowing which regions to focus on requires a technique that sees segments underneath defeating the difficulties and key activities to an effective convergence. An OT-IT convergence will be a basis for cutting edge supply chains, as it will empower the synchronization of information and data all through R&D, manufacturing, marketing, and distribution. Organizations with smart integrated technologies increase all-around in their business operation and stay ahead in their industry.

**Conquering the difficulties**: In any case, contingent upon industry, OT-IT change could be precarious or even troublesome. The horizontal instead of advantageous advancement of OT and IT implies that organizations face difficulties in completely relocating towards a consolidated OT-IT frameworks architecture.

**Crumbling hierarchy**: By and large, regardless of some current collaborations, OT and IT are housed under isolated offices, frequently with duplication. The merger will require an organized methodology, an operational development appraisal, prioritization of digital activities, and setting up of an advanced working process model. IT staff and OT teams bring diverse essential ranges of abilities and inspirations. OT employees have engineering or manufacturing backgrounds, and the skills in keeping machines operational. IT experts are enthusiastic about software developments, the power, and the process of programming. Bridging the divide between IT and OT can harness the power of enabling digital business, IT brings the speed and scale, while OT runs the business production and processes.

**Securing specialized interoperability**: Numerous organizations can report accounts of essential IT applications causing crashes in hardware or operational frameworks. OT frameworks themselves experience the ill effects of interoperability deficiencies; tasks equipment is frequently connected sub-ideally with various parts made for particular, confined capacities. Fashioners of Industrial IoT (IIoT) frameworks must take a stab at more prominent interoperability, both inside heritage activities innovation and in its grip of IT.

In summary, it is evident that IT and OT are essential features of technological innovations and the level of disruption that we have been discussing in this book, but all the benefits of IT and OT will be void if the end-users can't relate with the benefit. Unlike the previous chapters that highlighted technological transformations Inside-Out and Outside-In with respects to the business and organization structures (that will be discussed later), end-user centricity highlights how technological revolution and digital disruption has brought great value to business models by taking the perspectives of the end-users into consideration.

# CHAPTER 9

## END-USER CENTRICITY

As mentioned in the previous chapter, this is the modeling of the business plan around and for the customers. In this case, the customers dictate the extent and type of production to be done. While it is true that for any company or establishment, profit generation is the ultimate goal, for a company with a customer centric production, the best way to achieve this without betraying their interest, to make use of the available customer ideas, is to propel their production.

The central place of the customers is essentially the reported problems the end-users face with certain products which can be modified, reworked, repackaged and made better. The customers are not forced to consider the products or services as a second option because it is their complaints that birthed such products and services in the first place. Customer centric services ensure that you already have a booming market for the product as opposed to the product centric services that 'assume' that the product would be indispensable.

The difference between customer centric and product centric services is that while the former prioritizes the customers, the latter places more emphasis on the company's effort. The customer centric service, however, is more likely to succeed because of easier reception. The Pareto principle reveals that 20% of the company revenue is generated by customers. This proves that, especially for new businesses, the foundation of customers' needs must be a factor that decides the startup of the company.

Feedbacks are indispensable tools in maintaining the goal of rendering services based on the demands of the customers.

In a digitalized world however, the designs and concepts behind the presentation of an internet-based business should be done in a way that will be easy for the average consumer to understand and make use of. The key strategies to a perfect customer centric service include:

**Availability**: — The customers must be able to purchase the goods and services at their convenience and at any time. The customers would be more willing to buy from company A whose network services are faster and has reliable staff than from company B whose staff are unprofessional. The goal is to make the services truly available. This is in a dual format. The physical and the internet-based purchase should be active.

**Perfect delivery**: — Demand is one thing, supply is another. The customers should be able to get a perfect delivery service of their purchased goods. This means that the company must deliver to them at their doorposts as well as provide the qualitative products and services that were ordered.

**Personalization**: — The services and products must be made to suit the taste of the customers individually. By doing that, the company will entrench a marketable value in them — a feeling of belonging to the company and vice versa.

**Core capabilities:** — Regardless of how important the customers are; the workers of the company are much more important because they are like the magnetic field and can either attract or repel the customers depending on their 'charges'. A customer-based company would have customer friendly workers and a customer friendly atmosphere.

Most studies show that the customer experience will be foundational for the business transformation success. 56% of consumers will make their choice of products and services based on Customer Service/Experience. According to one study, a customer waited only 3 seconds for a web page to load, less than a year later, that was reduced to 2 seconds before a customer would move on, and do something else, or look for another service. This could mean a massive positive/negative impact on revenue and reputation.

The end-user centricity of technological innovation is the customer; they have control of today's marketplace. Customers realize that there are numerous choices accessible to them, few miles or phone calls away. With digital transformation, customers can perform more research and shop intensely; this is much simpler now than at any other time, more so why customers are smarter and anticipate brands and products to be easily accessible to them without much stress. End-user centricity, is much more than a word or a process, it needs to become a mindset and a culture of every organization. Organizations are undergoing digital marketing transformations, making a corporate client culture, and improving customer experience. End-user centricity is the act of concentrating on doing what is best for your customers consistently from both the internal and external environment viewpoint. Putting customers at the center of each business perspective makes end-user centricity completely proactive. Thereby, foreseeing client needs and the ability to divide markets into segments become easier and more effective. Customer centricity is a logic plan in which the requirements and desires for the end user of an interface are the focal point of core interest for organizations where they have to tune in and respond to client needs and measure client experience ceaselessly. Understanding the world through the eyes of the customer, and to continue doing what is best for the customer would keep organizations profitable and sustainable in the long run. Being a customer implies creating a long-lasting impression in the mind of clients that makes them resonate with a brand, learn, buy, contact, draw in and stay loyal to a brand. Putting customers first at the center of any business decision is the best way to stay profitable in a highly competitive business ecosystem. The best methods for achieving customer centricity include incorporating customer-centric marketing practices and efforts, by taking data-driven actions based on customer insights and building marketing and sales, around targeted customers.

## Design Thinking in Customer Centricity

Design thinking depends on the human capacity to be instinctive, to perceive designs and to build thoughts that are as genuinely important as they are practical. Design thinking uses components from the planner's toolbox like empathy and experimentation to arrive at imaginative solutions. Customer needs are settled based on future needs as opposed to recorded information or making unwanted analysis rather than trusted facts and trends. Design Thinking is a critical thinking strategy that helps organizations and people get an ideal result on an inward issue, or look forward to a future concept or plan. Design Thinking can include an immense business plan that ensures products are attractive for customers, yet is in tune to an organization's spending plan and assets. Design Thinking can altogether diminish the measure of time spent on planning and innovation. Design Thinking is tied in with testing suspicions and built up convictions, urging all partners to consider some fresh possibilities. This cultivates a culture of advancement which expands well past the market segment.

The principal phase of Design Thinking is called Empathy; it is intended to show signs of improvement to comprehend the issue that you wish to overcome. This incorporates connecting more into the issue to comprehend the current issue and having more profound cognizance of everything that is included with the issue. Empathy in business is becoming more acquainted with the customer and understanding their needs and challenges. This implies connecting with individuals so as to comprehend them on a mental and passionate dimension. Design Thinkers choose to understand needs with regards to current issues thereby putting aside their own suspicions. Amid this phase, in the design thinking process, comes assembling researched information. The Define stage will enable thinkers to accumulate incredible thoughts and have the capacity to see how to utilize them adequately. The second stage in the Design Thinking process is committed to characterizing the issue which is defining all data picked up amid the Empathy stage. When issues have been defined, design thinkers can begin the ideation stage. The Ideation stage considers an elective method to take care of standardized issues. Where design thinkers begin to utilize the data from the past stages to create logical ideas, and with a strong comprehension of customers' needs at the top of the priority, it becomes a great opportunity to take a shot at potential solutions. When the possible solution has been concluded, it is best to design a prototype by making various inexpensive products with a specific feature. With each new model, design thinkers should research how a prototype can meet customers' needs. The Design Thinking process is about experimentation and transforming ideas into the needed product. This progression is key in putting every arrangement under a magnifying glass and featuring any limitations and imperfections. Design Thinking can also be called testing where design thinkers test their models made in four stages of empathize, define, ideate and prototype in order to test models to perceive how well they can handle the current issues and needs.

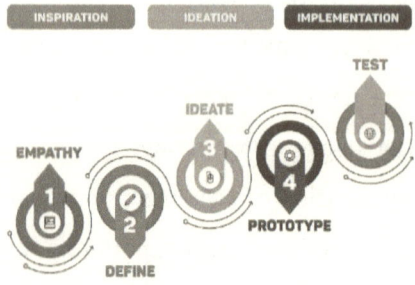

## CUSTOMER CENTRICITY INNOVATION

Customer-driven innovation is tied in with putting the customers first which could bring numerous benefits in any organization, to listen to the customers and put them at priority in end-user innovation design. Words are powerful, and they are how customers can create an impression about a brand. A happy customer can enlighten 2-3 individuals concerning their involvement with an organization while a disappointed customer will tell 8-10 people their ordeal with a particular brand which in the long run can cause a negative effect for business organizations. On social media platforms, like Facebook, Twitter and Instagram, a post or tweet can be shared to a great number of people like customers' social media influencers who have the power of reviewing certain innovation and upgrades across their social media platforms. Some social media influencers review any new gadget, movie and other products and give ratings based on their perception and customer experience. This is the reason end user centricity on the users of a product should be put into consideration by manufacturers, inventors, and organizations. For instance, some smartphones perform better than the others due to the singular reason of the innovation behind the technology. I walked into a phone store some time ago and a new phone was being launched and the population of those who were to test the product were majorly young people known as the millennials. These were the end users and possibly the market target. Customer-centric innovation has a great effect as it improves communication between the organization and its client base thereby enabling the organization to give an additionally fulfilling customer experience that can turn over a first-time customer into a reliable long term one that additionally gives a referral, and in the long run more benefit for the organization. Customer Centric Innovation gives an organization an edge over its rivals. Facebook enabled security features soon after the Paris attacks; the Safety Check gives users a chance to tell their family that they are protected in their various locations to avoid panicking and fear. Customer-centric innovation requires a continuous, sustained and focused effort from organizations in order for them to achieve the best result. According to Price Water Coopers global innovation report (2016), the world's leading brands grew by 35% due to their culture of continuous innovation in all their products which enabled them to achieve a record high of $680 billion in profit. These brands were able to apply the concept of continuous product improvement and innovation which enabled them to grow their market share and develop successful products. Adopting a customer-centric strategy to innovation makes leading brands sustain business growth because they have a deep understanding of end user needs thereby forming connections to customers and working around their yet to be verbalized wants. Innovation is a hot trend in the business ecosystem today as every organization intends to give the best value proposition to their customers thereby concentrating their research and development plans on customer-centric innovation.

Customer-centric innovation is important to business innovation because it is the core of the research and development process which is supported by customers.

End-user centricity innovation can be achieved through customer observation and field studies, analyzing data to uncover trends, and gathering direct input into the process to know the innovation that is needed in the next product upgrade and also to gain a competitive advantage in the market. Marriott hotel chain utilized an open innovation lab strategy to welcome customers to add to their structure and design for rooms, open spaces, meeting rooms, and food ideas by allowing them to test and bring in their suggestions on the best innovation for modern hospitality service. This prototyping strategy gave Marriot unparalleled moment criticism, which was utilized to execute developments rapidly in its properties around the globe and today Marriott International is a leading diversified American hospitability company which has a history of more than 80 years of delivering the best services in the hospitality sector. Innovation is all about how improvement drives consumer loyalty. Feedback from customers is the best source of incremental innovation for brands that desire continuous growth. For instance, US-based financial services firm Merchant Cash & Capital (MCC) accumulated gathered customer criticism through their online surveys and personal interviews. Loan officers learned that the customer onboarding process was inconvenient, with a slow and inconvenient process that affected end-user satisfaction. The company quickly engaged their technology partners to digitally transform the process of their financial service, by making loan applications very easy to complete with just a few clicks for any customer. In this manner, they were able to improve on their general customer experience.

Customers are not just connected through new gadgets and applications; they are likewise increasingly social and have more distinct needs. Businesses need to be well positioned with the correct message or offer, at the ideal time without fail in the physical and virtual world to connect with today's customers and earn their trust and loyalty. With the end goal for this to occur, brands should carefully change to meet the advancing needs and desires of their customers by improving the customer experience. Customer centricity and creating positive experience encounters should be a priority for businesses who want good company image.

## DIGITAL TRANSFORMATION FOR CUSTOMER-FOCUSED STRATEGY

**Omni-channel**: The present customers hope to have the option to buy products at any point and any place they wish to. As indicated by Forrester, organizations that offer customers the capacity to buy through both physical and online areas have 30% higher lifetime value than those that only utilize one channel.

**On-demand request**: Customers likewise anticipate that goods should be delivered to them as they wish. They anticipate that requests should be merged and delivered to be packaged with additional services, thereby making things faster and much easier for them.

**Personalization**: Making a customized experience depends on customer information gathered. This one is sensitive because enquiries on customers have to be made seeking permission from customers in order to utilize their information and meet their special needs.

**Core capabilities**: To improve client experience, it needs to start within an organization. The front, center and back office must be fully connected to meet customer needs. The organization can live or die by the way it delivers services, responds to issues, and manages customers' expectations. Customers, employees, products and services contribute greatly to the customer experience.

## PRODUCT-CENTRICITY VS. CUSTOMER-CENTRICITY

A product-centric organization is basically concerned with product dominance, a methodology that is driven more by research and technology innovation. A customer organization centers on the diagnostics of business issues and offers some benefit through altered solutions. It is an 'outside in' approach driven by innovative service delivery experience to satisfy the customer's enthusiastic needs. The essential objective here is long term relationship, and it goes for the mind offer of customers instead of a piece of the overall industry. Customer centricity takes a look at buyer-driven approach as versus sales-driven approach. Apple is an extraordinary case of a product-driven organization. In Steve Jobs' words, "Customers don't have the foggiest idea what they need until you show it to them." This thought is at the center of Apple's organizational and market structure that considers the best and most beneficial delivery of their exceptionally innovative products to its customers. Amazon is at the furthest edge of this thought. Jeff Bezos' e-commerce philosophy centers on this thought, "If you're competitor-focused, you have to wait until there is a competitor doing something. Being customer-focused allows you to be more pioneering." Everything that Amazon does rotates around customer centricity, to make them happy always, because the customer is king in business. The Digital Customer is King. The financial aspects of drawing in and holding customers show why organizations place major emphasis on customers as opposed to product centricity at the focal point of their tasks. Research on customer psychology indicates that losing 5% of new customers can increase profits by as much as 75%. In terms of price sensitivity, existing customers are much more forgiving than new ones.

# CUSTOMER-CENTRICITY AND DIGITIZATION

Digital transformation (as discussed extensively in chapter 2 above), at its very center, is extremely about customers and their encounters. What's more, the way to advanced change is out there for all to use. Rich customer information is separated and covered up in various storehouses across the organization. Taking advantage of this and coordinating it for insights and logical activity can put customer involvement in the middle to make each touch point a significant one. Content optimization, in the cutting-edge advertising world, is key. For the millennial or Gen Z buyer, improving contents for business visibility becomes more significant. Lead generation can come by rendering ads in a format that is native to the user's device or with the content of specific interest to this segment. Digital and social media platforms certainly have the ability to track and share user behavior extensively for better content and channel activation strategy. Social listening could be utilized across various social media platforms to get feedback on product reviews, brand reputation and also, customer complaints. Digitization accordingly empowers a 360-degree bound together view from numerous information sources to touch base at well-characterized client portions and tweak focusing on procedures. Globally we see organizations embracing digital changes with the innovation and assets that are accessible in the particular country so that they keep meeting the needs of their customers. Customer closeness enables organizations to avoid the commotion and mess of rivalry; the customer is the main course to progress. Customers are the major source of unlocking present and future trends in business ecosystems. Collaborations with industry influencers is the best way to advertise in today's digital world because these influencers have a lot of followers across their social media platforms and can easily generate leads for organizations through their social media accounts. Social media influencers have the ability to promote products by encouraging their followers to patronize and remain loyal. In case you're selling inside the Internet of Things space, who are the bloggers covering the subject? Regardless of whether you're utilizing LinkedIn or social listening devices, you ought to search out idea pioneers, experts and early adopters. Conventional strategies aren't working, as buyers go to the internet to search for what they want, the best way to get them to buy is having the best content and digital marketing strategies.

## END-USER CENTRICITY IN DIGITAL AGE

Technologies like mobile devices, social media, the cloud, the Internet of Things, and artificial intelligence (AI) are rapidly changing customer behavior and also enabling organizations to customize customers' needs in an entirely different manner. E-commerce sites like Amazon have customers purchasing history and recommendations for customers to keep them in trend with the latest products on their website offering discounts and gift cards to enable more purchases. According to the sales force, technologies of the Fourth Industrial Revolution such as artificial intelligence and the Internet of Things are enabling businesses to offer better customer experience because expectations from the customer has greatly increased over the years. Organizations must give outstanding customer experience to remain in the game of the competition. Customer centricity in this digital age means providing exceptional service before and after sales with customized customer experience across online and offline channels. A remarkable number (84%) of customers state that being treated like an individual, rather than statistical data, is imperative to winning in business. 76% of consumers and business owners say that it is easier in this digital age to move their business to any location once the customer experience is done effectively.

Co-founder of Apple, Steve Jobs, knew the significance of customer centricity when he said the first step is to begin with the customer experience and innovation follows after. Since customer experience keeps running on information, building trust with the customer in this digital era is crucial. Information ruptures have been increasingly more frequent, with several online businesses at the risk of cyber-attack. 59% of customer's trust their own data is defenseless against a security break, while 54% of customers have zero to little trust on organizations having their best at heart. Information abuse has left numerous individuals considering how their own data is truly being utilized. However, 93% of customers say trusting a company makes them more likely to recommend that company. Today, social media and review websites make it easier for customers to give their review on a brand to a broader audience than in the past. The advantages that organizations that gain customer trust have is numerous. Customers spend more with brands they trust, and they are also bound to recommend them to their family and friends. Organizations must become more customer-centric if they want to survive in this Fourth Industrial Revolution. By making customers the core of each collaboration, customer-driven organizations can make a dynamic, consistent, and extraordinarily customized involvement for customers to accomplish goals, and achieve ideal results, and eventually this will drive higher sales and return on investment.

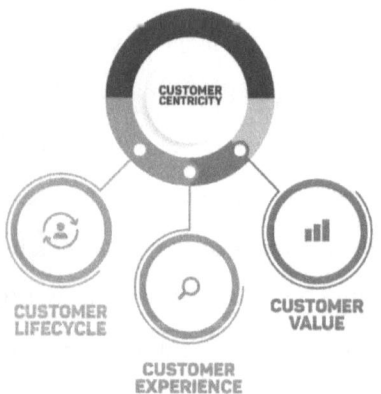

# TECHNOLOGY ENHANCING CUSTOMER EXPERIENCE

Technological advancements over the years have given rise to smart customers who have grown to expect higher standards of excellence. In the age of digital transformation, innovations such as virtual reality, cloud computing, big data, live video connectivity and intelligent Chabots are changing the manner in which organizations interface with their customers. Remaining with these advancements and patterns will include acing these new developments and utilizing them to deliver, connect with and advance in innovative ways of doing business. To remain outstanding, organizations would need to embrace technological innovations and go beyond what their competitors are offering customers, with improved user experience. Technology is vital in bringing ideas into reality and giving customers seamless user experience, therefore, a technology that is ineffectively executed or requests a lot from the end user can truly overrule the improved user experience.

According to the information age, the following technologies are enhancing customer experience today. Chabots are the most recent insurgency in the business scene. With the assistance of AI-controlled Chabots, organizations are better prepared to deal with customer service support. These menial helpers improve customer relationship management. Chabot enables businesses to provide instant support via voice, mobile app, instant messaging, SMS, or websites as these bots guarantee brisk reaction times, client questions are taken care of productively, improving consumer loyalty and involvement with a business. Big data analysis is also a technology helping to improve end-user experience, as indicated by a review by Salesforce. 57% of online customers readily share individual information with organizations that guarantee to send customized offers and protect their personal information. With enormous data gotten by a business through different platforms, it is progressively significant for them to investigate this information and use it for conveying a superior encounter to end users. Smaller organizations can now assemble valuable bits of knowledge into buyer inclinations and practices that would help them give customized services. The idea of personalization enables a business to assemble a dedicated client base and improve incomes over the long run. Big data is changing the manner in which organizations communicate with clients. The innovation helps a business understand clients' issues, and needs. Virtual reality is the latest innovative leap forward that is changing the buyer experience scene for current organizations. VR can draw in clients in a superior manner, as the innovation is intended to give a total physical encounter that catches the consideration, creative energy, and faculties not at all like some other innovation present in current time. Offering a vivid encounter, VR is helping organizations to draw in shoppers in a superior manner than different innovations, spurring buyers to connect in a superior manner with brands that entice them.

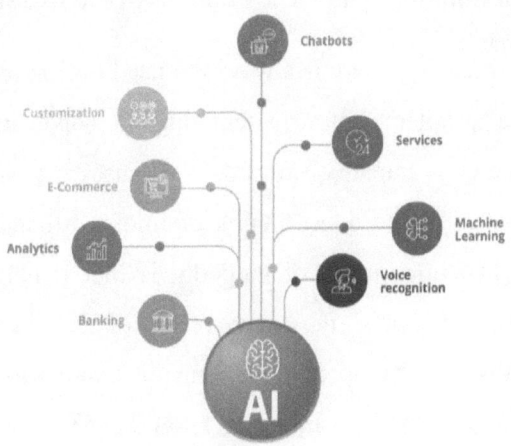

## DESIGN THINKING

Design Thinking is the process of thinking out logical solutions to human needs and designing or providing solutions to the problems after a series of brainstorming, ideation and prototyping. Design thinking is an art of the logical and industrial mind that is needed to get out of tight situations. It is a cognitive, strategic and creative process as well. Design thinking is important in the digital world to minimize chaos and answer questions that defy linear solutions.

**Design thinking is a process that occurs in five stages:**

**Empathize**: — There is always the need to first identify with a problem before attempting to solve it. The absence of empathy towards a cause means that automatically one doesn't think of finding a betterment to it. The importance of empathy is that it helps the designer to know the very reach of the problem. Empathy here has to do with want — wanting to know what is happening, wanting to know how terrible it is, wanting to know what can be done.

**Define**: — The next step into a successful design thinking process is to define the problem or the need of the user, the target of the design and how much it will cost. There is a need to evaluate the why and the how much, in the quest to reorder a disrupted place. This would help to prevent some bad risks and replace them with certainties.

**Ideate:** — The third stage of design thinking tasks the designer with creating something out of nothing. Bringing out the solution right from the middle of the problem. Brainstorming process could take months and could take minutes, the designer in this stage needs to bring out the daring alternatives and the out of the box approaches that are most needed. The ideas must not be in conflict with the written definitions.

**Prototyping**: — The designer needs to employ the ideas in stage three in a small-scale format and observe the results to decide whether the ideas are good enough or not. The designer must be sure of the results before employing them in a large scale, in order to prevent huge losses. A failed prototype might mean that the designer must make another prototype until it works or revert back to stage three to remake his ideas.

**Trying the solution:** — Once the designer has gotten a perfect prototype, the next item on the agenda is to use the same solution in a large format that fits into the customer's reality as an answer to a question.

Having discussed end-user centricity or rather still customer centricity, what about those making the work happen, those driving the innovations in the products and services offered to the end-users? Yes! I am talking about the workers, or rather the employees, they are the ones that implement the feedback from customer centricity to help provide better products and services to the end users. How is technology transforming how employees work? So many tech startups go with the hashtag that the future is now; how is the future of the workplace looking like? What needs to be factored in to make things better for employees? All this and more will be discussed in the next chapter (Employee Centricity) where we will be breaking down the level of technological disruption in the workplace.

# CHAPTER 10

## THE WORKPLACE OF THE FUTURE

Employee Centricity is the opposite of customer centricity (as discussed in the chapter above) or the product centric service rendering is the **employee centric** based service. This type of business is run by making the employees the biggest influencers. They are given to a considerate extent, fair treatment, and good working conditions. What are the consequences?

Employee centricity breeds a perfect working environment. For someone who wants to run a business in a peaceful and harmonious manner, it is advisable to consider being employee centric. The chaos and misunderstanding that often characterizes certain businesses would be removed in a company where employees are given the freedom to do their jobs. This would in turn give birth to a smooth flow of work and production would be easier.

The company would also be able to leverage on the insight and talents of the workers. In some cases, the company might need to depend on the intuition and insight of the workers to complete projects and production. The workers in an uncomfortable environment would want to hoard their knowledge and insights in such cases, whereas in a comfortable place, they would readily offer their personal assistances in difficult situations.

Interaction with customers would be easier, and necessary feedbacks would be received. In a happy place, the employees would not only satisfy their employer but also the customers. This would in turn boost the image of the company and facilitate success in the long run.

Many companies say that their biggest asset is, their employees! However, when it comes to reality, how many companies put the employee first when it comes to implementing the strategy (Inside Out)? Just to be clear: Building the strategy, OUTSIDE IN, customers come first. Implementing the strategy, INSIDE OUT, Employee comes first.

The development of the work environment has seen much advancement since the era of digitization. The evolution of workplace has taken an exponential turn since machines were introduced during the first industrial revolution. The universe of work is evolving, a portion of the procedures we rely on in the workplace is gradually vanishing and new technologies are advancing to replace traditional workplace. How work is completed, by whom and where are for the most part changing at a remarkable rate. The main reaction is for individuals, and organizations, to adjust to these changes, adapt to them and grasp them. Organizations need to remain on top of things to stay focused, and workers should be eager to be champions, as opposed to obstructions, to change. New advancements and better approaches for doing work will shape a future that puts individuals first, making work increasingly important. A portion of these progressions is bound to happen; real moves in the work environment that put an emphasis on human ability and affect how human resource management (HR) would work is on the rise. The transformed workplace providing and facilitating collaborative teamwork environment, is becoming extremely valuable in the future of work; while video, mobile, and social, are increasingly becoming tools for the virtual environment. With the rise of AI, chatbots are making an appearance in numerous work environments, and soon, organizations will most likely depend on them to deal with authoritative assignments so they can concentrate on progressively significant work. There would be an increment in Data-Driven Decision-Making. Organizations will most likely actualize wise work. This implies HR will almost certainly invest more energy in high-esteem work and less time on authoritative assignments. As email diminishes, inefficient gatherings vanish, chatbot gets received, and information becomes increasingly usable, workers will almost certainly invest a greater amount of their energy becoming more creative and highly skilled in several fields.

The possibility that robots will assume control over our employment is probably not going to turn into a reality. Rather, people and AI machines would cooperate in making the future of work more productive and efficient. People will dependably be better at fundamentally considering, teaming up with colleagues, and working virtually with people across the globe just to foster collaboration in the digital era. The future work environment would enable workers to invest less energy in everyday repetitive tasks and assignments because creativity and collaboration would be the future of work. All work done by people will be imaginative. Laboring through unlimited spreadsheets, dealing with many messages, sitting idle at work and being overpowered with authoritative errands will never again be heard of in the future. It is difficult to envision a more digitized world than the one we live in; however, the future working environment will bring collaborations and friendship in the real and virtual world. With the manner in which innovation will be actualized in the future of the working environment, straightforwardness into one another's work will build trust and sustain working relationships on the long run. Workers will become much more open with one another which would make work more productive and faster. Technology advancement would make work in the future easier and faster. No more 9-to-5. The workplace has, for a long time now, remained the center of business. A space where people assemble for a few hours every day and focus on individual undertakings. While this idea of the 9-to-5 workday is still to a great extent prevalent, frames of mind are evolving and this traditional system will soon become a thing of the past as digital transformation will change the future of work gradually.

## EMPLOYEE CENTRICITY

Clearly, there are numerous angles an organization must consider so as to attempt to improve working conditions for employees. Encouraging employee participation would boost employee creativity and performance which, in the long run, would make an organization profitable. A system concentrated on the worker is pivotal to urge employees to add to the business execution targets. A worker is at the focal point of organizational objectives when they are allowed to contribute and see how their individual and aggregate commitment can help shape the business, obviously, when this is normally and reliably exchanged through straightforward procedures that create trust and certainty. At the end of the day, estimating and imparting to workers the effect everyone has on the business is a basic factor for an employee-centric strategy to take shape. Building and securing a positive culture can feel overwhelming to HR pioneers, particularly as an organization develops by hundreds, or even thousands, of workers. HR needs to build a solid relationship between employees and management executives in order to achieve business objectives successfully. Every single incredible organizational culture requires a pattern of compassion and human comprehension to really flourish, where every worker in an organization is valuable and regularly appreciated for their efforts. Trust is built when workers are in the know of the situation of their company prosperity. The organization that celebrates together remains together. Workers like to get together and relax in cheerful work environments; it feels better, and it carries positive vibes into the workplace. An organization that has truly invested in an employee-centric approach on its workers can hope to see its organization vision enlivened. Workers will feel a certified want to contribute towards the accomplishment of the organization's mission and experience the organization's admirations each and every day. Just when workers feel sure about their mindfulness and comprehension of procedures, will they have the capacity to put in their best at the workplace? An employee-centric approach is devoted to putting the experience of workers first because they are the first internal customers and their productivity would affect all areas of the business. Workers' welfare is linked to business performance in every aspect of business operation.

## STRATEGIES TO EMPLOYEE CENTRICITY

Good welfare — Every employee deserves good welfare, not only because they must be in good health to be able to discharge their duties, but also because they have contributed and will continue to contribute to the growth of the company. Good welfare must be one of the basic strategies of an employee-based production.

Conducive environment — The employees must have the needed tools and machines at their disposal to prevent hitches in production and also to save them from frustrations. Also, a friendly boss isn't too bad for an employee centric system. Praising a worker is a much more effective tool to enhance productivity than criticizing him/her.

Training and Drills — The Company must have regulated programs that will serve as a sharpening and reorientation for the workers at regular intervals to prevent recycling outdated information and techniques. Special courses and seminars would go a long way to help the employees.

Career growth — A company that wants to get the best out of workers would monitor their careers and ensure that they do not just get better at their work, they also get better at their education and on a personal level more so. The workers would appreciate a company that allows them to spread even more and progress in their respective careers.

## THE GIG ECONOMY

Digital transformation will change the course of some industries, and will definitely bring some people into the unemployed circle. The answer to the question of how can the economy sustain the unemployed but large population, is gig economy. Gig economy is another evolving process that the world is going through. This transient job will put an end to traditional employment in some industries in the coming years. Some of the reasons that gig economy is embraced are:

Flexibility — It is easier in the gig economy to do different jobs ranging from one extreme to the other. The availability of temporary jobs makes it easier for workers to stretch their mental soundness as well as their physical state. The jobs are easier than most traditional ones and can be completed in a shorter time frame.

Decentralization — the gig economy is regarded as a free market where no government can impose certain conditions of work. It is a classless e-workplace where there are no rules and workers are free to make choices and earn their living.

Internet based — Gig economy doesn't require a physical presence before work can be done. This gives it an edge over traditional employment of 9am to 5pm. And besides that, the gig economy is more profitable. Outsourcing has been a thing for a considerable length of time. However, lately, the number of freelance professionals has greatly increased. Millennials place a higher incentive on flexible work which is giving rise to the Gig economy and which incorporates general workforce conditions in which there is financial commitment, contracts, labor and deadlines. The gig or independent economy varies from customary work in that employments are not permanent. The gig economy works on innovation stages that mean to associate skilled professionals searching for adaptable work game plans with the organizations who need them in a concentrated area. Specialists looking for adaptable, transient working game plans and organizations trying to procure outsourced labor have given rise to the gig economy. In the independent economy, the nature of work is self-employment, which means customers pay based on the agreed rate for jobs rendered. People who are signed up to platforms on the gig economy can make a living with the right skills and strategy. Individuals from the workforce with full-time professions, who need to enhance their pay, can, without much of a stretch, get a couple of gigs in the nights or on the weekends. Gifted experts can apply more power over their vocation by taking part in testing tasks and building a noteworthy profile that would enable the gig to be among the best freelancers. Today it is simpler than at any time in recent history for organizations to use the gig economy as an outsourcing strategy. The greatest advantage of the gig economy is adaptability; deciding your accessibility early makes it simpler to remove gigs that work with your timetable and lessens the probability that you would have to make changes or bow out of a formerly acknowledged gig. The gig economy offers numerous open doors for the workforce and managers alike. The rise of the autonomous workforce is digging in accelerating the acceptance of the gig economy, because of innovation progressions that make it simpler than any other time in recent memory for skilled freelancers to secure impermanent positions in which they could earn while being flexible and productive with their time. The Bureau of Labor Statistics announced that 55 million individuals in the U.S. are "gig workers" which is over 35% of the U.S. workforce. That number is anticipated to hop to 43% by 2020. Gig work is essentially a way of depicting an autonomous contract or low maintenance work, such as driving for Uber or independent copywriting. Twenty to thirty-year olds are rapidly floating towards gig work for the guarantee of a more noteworthy work-life balance. Boomers and different ages on the very edge of retirement are attracted to gig work since it acquires some additional pay without a noteworthy time duty. The gig economy is extraordinary for creative youth. When you are a youthful innovative simply beginning, and your resume is pretty much a clear slate, gig work can enable you to get a foot in the entryway. The gig economy permits creatives to pay the bills while giving them an opportunity to seek after their interests.

Fifty years back, the standard of work was getting your first in a steady organization, check in and out for quite a few years, and after that resign; there was no need to stress over when the next paycheck would come. But today it is a different ball game as a result of the digital transformation that has swept through all industries. Today, millennials go into the workforce realizing that their profession will probably be in motion. They must be set up to continue learning and foreseeing patterns so they do not end up surprisingly out of work. A lot of gig workers begins their vocations by bouncing on a venture on the grounds that the business is urgent and needing assistance now. In any field, climbing the positions requires a great deal of long periods of training and development. It may seem hard getting the first gig because the pool of ability is brimming with gig workers who have not been allowed a chance to sharpen their skills. Organizations need to develop to figure out how to represent a flood of impermanent specialists. This implies managers must put resources into mentorship and training. When organizations figure out how to flourish with gig specialists, everybody wins. The pace of progress in the worldwide workforce is quickening. To succeed, we need to adjust similarly as fast.

Since the term gig economy was popularized around the height of the 2008-2009 financial crisis, task-based labor has evolved and has become a significant factor in the overall economy. The concept of creating an income from short-term tasks has been around for a long time. The gig economy is very broad and encompasses workers who are full-time independent contractors, people who are driving for Uber or Lyft several hours a week. In some cases, the worker is a small business owner, and in others, they are freelancers who are paid to complete discrete projects for larger organizations. Musicians, photographers, writers, truck drivers and tradespeople have traditionally been gig workers. The term gig arguably came from the music industry, in the argot of jazz musicians, attested from 1915. The gig economy suddenly became much more than a curiosity during the aforementioned financial crisis. With swaths of the population facing unemployment or underemployment, many workers picked up temporary engagements wherever they could. These gigs had to be flexible. Some workers were able to hold down a full-time or part-time job but needed to shore up their income. Others cobbled together an income by working a few gigs at once. A sizeable portion of the economy is driven by technology. Software platforms that enable the sharing economy and gig economy are a successful business model. Gig workers do not have human bosses, they work for apps. The market is becoming richer and more complex by the day. Perhaps more than ever, we need opportunities for new businesses and new entrepreneurs; opportunities that do not require access to deep capital. The sea change we are seeing in the global economy brings uncertainty and insecurity, which could be well addressed by opening up common networks in the big verticals, to encourage innovation and independence within the labor force. Task-oriented work has already gathered tremendous steam, which suggests that in the nearest future we would see an economy that has rebuilt itself on hundreds of millions of small businesses rather than hundreds of millions of 9-to-5 jobs. Then the gig economy will be the most reliable and productive platform that has created lots of employment and connected people globally.

## HR TRENDS AND PREDICTIONS

There will be a keen interest in ways to improve employee productivity and well-being, and the rise of AI-driven HR (Human Resource) technology will make the HR increasingly coordinated and productive.

**Employee engagement**: G2 Crowd predicts that organizations will build their representative commitment employee engagement spending 45% from the budget by the end of 2019. When employee commitment is low, organizations are at a great disadvantage. Organizations need to do more to enhance employee welfare if they must survive this digital era. It is very easy for a dissatisfied employee to go online and rant, thereby creating a bad reputation for the organization.

**Use of technology to remove unconscious bias from hiring**: The main test is that more often than not, HR faculty have oblivious predispositions as far as sexual orientation, race, nationality and so forth. To bridge this gap, G2 Crowd* predicts that organizations will expand their utilization of innovation to expel procuring predispositions by 30% in 2019. (*G2 is an online peer-2-peer review platform).

**Corporate wellbeing activities**: Corporate wellbeing activities concentrating on budgetary, physical and psychological well-being are relied upon to rise by 40% in 2019. These solutions are intended to support worker confidence and improve efficiency.

**Utilization of Artificial Intelligence in HR**: Human resource (HR) keeps on improving. What's more, in 2019, G2 Crowd predicts that AI-driven HR tech will increase by 35%. This is, for the most part, because of the positive improvements in AI and the re-established concentration to place HR in an increasingly dynamic job in the work environment.

**The trust issue**: An ongoing overview directed by Ernst and Young demonstrates that not exactly 50% of employees have "a lot of trust" in their present bosses. This is disturbing as most HR activities are planned dependent on the idea that employees trust the organizations in which they are working, which happens to be contrary to what organizations think.

**The human touch**: Artificial intelligence will upset the manner in which we work. What's more, in opposition to mainstream thinking, it will create more jobs than it will replace. While AI is chipping away at monotonous and routine errands, we will have additional time that would enable us to excel at other creative and significant tasks.

**The rise of social capital**: There will be a ton of spotlight on social capital-boosting which is the estimation of how people and groups are connected together in the association. This development is led by organizations that utilize Organizational Network Analysis (ONA) to feature key influencers and distinguish burnout dangers.

**HR innovation showcase**: High-tech companies are presently bouncing into the HR tech. Since 2009, $16 billion dollars have been put into the HR tech, and this is relied upon to ascend as Microsoft, LinkedIn, Google, Amazon, Oracle, and even Facebook are building their very own HR innovation.

**The gig economy will keep on flourishing**: The gig economy keeps getting solid. In any case, this isn't (yet) the eventual fate of work. Indeed, there are organizations that need to transform their gig specialists into ordinary employees.

**Work-life balance**: Employee work-life balance is significant. In spite of this fact, employees are unofficially requesting for adaptable methodologies that will enable them to balance their jobs with other aspects of their lives.

## THE SOCIAL AND ECONOMIC BENEFITS OF A GIG ECONOMY

Workers in the gig economy are normally independent or temporary workers. A key benefit of the gig economy is the ability of freelancers to control their time. This might be one of the key individual advantages. Investopedia depicts a gig economy as one in which "impermanent, adaptable occupations are typical, and organizations tend towards hiring freelancers and consultants rather than full-time employees". The gig economy may likewise lift individuals out of the underground economy by giving them real and adaptable work openings with low entry barrier. A freelancer in the gig economy is allowed to deal with their own yield, expectations and income. They have no commitment to work when they can work from any place, at whatever point, and for whomever they pick without restriction. Utilization of gig work platforms has developed by over 30% in rising economies. About 1-4% of freelancers in developed economies depend on gig platforms for work. The expansion is considerably increasingly stamped when you think about the people who are utilizing gig work to enhance their full-time pay. This could help organizations connect to freelancers from various countries, so as to meet their work needs. The gig economy is multifaceted to meet all work-related purposes in each industry. Regardless of fears of abuse and low wages, most gig workers wouldn't pick a long-term job. About 45% of them would pick short term independent jobs, and try to secure new customers so as to meet their financial needs. Gig economy pulls in people with diverse professional experience. 40% of organizations expect that gig workers would later join their workforce as a result of their level of experience. The development of the gig economy can be viewed as something worth being thankful for, whatever length of time that everybody is included is mindful so as to guarantee that the most unskilled workers aren't misused.

Much as we have discussed how technology is really disrupting the workplace and employee centricity, these impacts are mostly more reflected on the business or organizations that are providing these products and services. With either customer centric or employee centric, they both transcend to the impact of digital transformation on business as a whole, which leads us to analyze how to get involved with creating the impacts on digital businesses in the next chapter.

# CHAPTER 11

## CREATING THE IMPACT (THE DIGITAL BUSINESS)

Technology innovation has undeniably affected how we complete our everyday schedules. Technology has been adjusting to the requests and needs of every era in human existence. Technology is a living revolution that has caused digital transformation in all industries by fusing core data with machine learning to advance people, processes and organizations with seamless devices that help connect the physical world with the digital world. There are several brands like Amazon that have made an effective business from their online platforms. Utilizing GPS-driven automatons, Amazon can convey pretty much each and every item straightforwardly to wherever by checking deliveries with constant information from its applications and workers. Customer needs at all stages have become more successful because they are creating an impact through their digital platform. As innovation propels, it upgrades and improves the connection among brands and buyers and settles on better advertising choices dependent on target responses. In this age, it is essential to adjust to the quick business changes by completely adopting digital strategies in business operations. The inability to do so can scrutinize a brand's survival. The present-day smart customers are searching for dynamic brands that can bring together rising advancements with their current business procedures. By broadening the scope of business operations, organizations need to concentrate on ways to coordinate these new and conventional strategies together to work as one. The capacity of the organization to redesign its procedures and systems relies upon an unmistakable plan for future trends. Numerous organizations have just started to change themselves towards digital transformation as they realize that it is not something to be left for tomorrow. The developing availability of individuals, machines and even organizations has changed the requests of the business sectors. So as to keep up and remain aggressive, businesses need to change in accordance with these requests by digitizing their procedures and plans of action. Be that as it may, the advanced change likewise holds a great deal of new chances to develop or even set up new parts of the business. Along these lines, organizations should grasp advancement and guarantee powerful customer commitment.

## DIGITAL BUSINESS TRANSFORMATION

More people today interface with the web through their smartphones as opposed to decades back when most people used computers. Nowadays, organizations are building their product offerings to be centered around digital gadgets and mobile applications to keep customers connected and loyal. The foundations of digital business transformation can be traced back to 1995 from MIT scholar Nicholas Negroponte's book "Being Digital" which investigated the substitutability of bits and particles. Negroponte proposed that any type of data that exists as molecules like books and DVDs can be spoken to by bits on a computerized gadget. This further shaped the premise of the early development of web-based business. Since then, organizations have widely implemented digital innovations and adjusted procedures to be using advanced innovations and to tangibly improve execution. Digital business transformation is about smarter performance by connecting technology, organizations, people and trends to create a long-lasting impact. Social media platforms like Facebook, LinkedIn and Twitter take into consideration a two-route stream of data and communication between an organization and its key partners by enabling a sharing economy in which individuals are urged to express and utilize their inner and outer information for their trusted brands to improve customer experience. This has made positive impacts on customers and organizations. Digital transformation is quickly turning into a top need for organizations in each industry. The present organizations work in a universe of expanding intricacy portrayed by higher information volumes, quicker business cycles, and unending customer needs. In this unstable atmosphere, conventional plans of action, procedures, and ranges of abilities can put an organization at an aggressive weakness. Numerous organizations are setting out on an undertaking of advanced change to enable them to drive income development, upgrade the customer experience, improve efficiency and take advantage of opportunities. Digital transformation enables organizations to build up more liquid, responsive, and information-driven procedures that are essential for progress, where organizations can greatly utilize time, materials and resources more effectively. As organizations utilize advanced change to become increasingly liquid and responsive, they are likewise requesting that their workforce approach everyday obligations in new ways. The advanced economy requires crisp dimensions of understanding, imagination and adaptability from specialists. New technologies presently free the workforce from a considerable lot of the conventional manual and work concentrated undertakings that regularly occupy groups from increasing the value of the business. Today, advanced change is a notable idea. The developing network of individuals, machines and even organizations has changed the requests of the business sectors. So as to keep up and remain focused, businesses need to change in accordance with these requests by digitizing their procedures and plans of action, creating the value matrix, transformation at all levels Customer & Employee experience, and operational technology excellence. See figure below.

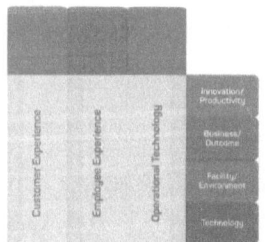

## DIGITIZATION BENEFITS

For every industry, digital transformation comes with an offer: that they may either take the lead in disrupting their own business model and design their destiny by accelerating the pace of innovation and transformation, or wait and face your demise. For those who choose to embrace digitalization, the benefits include:

**Digital nearness**: This is presumably the most obvious favorable position. The nearness on the Internet, through devices, for example, online stores, informal organizations, websites, corporate pages, and so on, duplicates the visibility of the organization and sales channels. For certain organizations, this nearness is the focal point of their digital technique, and they have even moved their business from conventional structures to online sales.

**New contact channels with customers**: Digital nearness opens up sales channels and new ways of communicating with customers such as email, applications, and social media platforms. These days, the customers of any business with a digital nearness have various medium they could use to stay connected to their favorite brands.

**The client at the core of the Universe**: The nearness to the customers infers this obligation. In the meantime, consistently, we have new innovative methods readily available that offer us new alternatives to improve the customer experience.

**Better leadership**: The digitalization of business makes it conceivable, as we have found in the past passages, to have nonstop contact with customers, and this enables us to become more acquainted with him/her. Be that as it may, that isn't the general purpose. A few organizations go further and apply Big Data when settling on a wide range of choices that influence nearly the whole business. Digitalization of business can prompt a huge increment in efficiency and can decrease a few expenses. Innovation has helped organizations improve in these zones consistently. Digitalization can likewise do this.

**Reduction in cost** — For the regular customers of grocery stores or constant travelers or book readers, digitalization seeks to reduce the cost of purchase by making it possible to purchase directly from the producers. Even for those in the industries, getting to customers becomes easier and cost is reduced as a result. Uber for example makes it easier for taxi drivers to locate their passengers with precision.

**Meeting customer's demands** — digitalization made it possible to have a satisfied customer. The demands of the customers can be met with speed and accuracy when digital means are used for transaction or transportation and other means. Digitalization has reduced the risk of displeasing customers.

**Fast adaptation to market conditions** — Unlike the pre-internet businesses, digitalization is ensuring the producers and consumers of near zero glitch in market value because of easy adjustments and adoption of up to date market conditions.

With the many benefits associated with digitalizing business processes to maximize productivity. The right framework and enterprise architecture must be put in place in order to be aligned with the forces that are responsible for digital business transformation. What are the essential enterprise architecture frameworks necessary to have a successful digital transformation in the business world? The next chapter will be discussing this framework in details.

# CHAPTER 12

## ENTERPRISE ARCHITECTURE
## FRAMEWORKS

Enterprise Business Architecture will be extremely critical for the successful business strategy and roadmap; therefore, this chapter will provide all details around the EBA.

Enterprise Architecture (EA) is the act of investigating, structuring, arranging, and actualizing enterprise analysis to execute on business techniques or strategies effectively. EA enables organizations to structure IT methodologies in order to accomplish business results and to remain over industry patterns and interruptions, utilizing design standards and practices. Enterprise Architecture enables businesses to accomplish their key objectives. EA helps organizations in developing competitive advantage, decrease risks, and improve productivity and versatility. EA improves the structure and operation of an organization; the aim of EA is to decide the adequate ways to accomplish present and future targets. Tasks are easier to achieve because IT infrastructure is aligned with business objectives; these techniques boost digital transformation, IT development, and the modernization of IT as an office. EA improves leadership, versatility to changing requests or economic situations, end of wasteful and excess procedures, and minimization of employee turnover. Enterprise Architecture Management improves organization's procedures, frameworks, and innovations by building up a comprehensive, interconnected model of the enterprise that incorporates forms, data, applications and innovations that provide a path from the current state to the target state or roadmap to future plans. Digitalization has played a major rule in Enterprise Architecture, as organizations stream over big business solutions to extend its clients and environments. The IT Standard for Business exhibits a methodology that gives more space for organizing the digitalization by coordinating business to various environments. The seclusion of this methodology permits an increasingly active improvement of design for various business zones.

EA started during the 1960s, and was conceived from "different structural compositions on Business Systems Planning (BSP) by Professor Dewey Walker", as indicated by the Enterprise Architecture Book of Knowledge (EABOK). John Zachmann, one of Walker's understudies, detailed those reports into the more organized organization of EA. The two men likewise worked for IBM amid this time, and that is when Zachman distributed the structure in the IBM Systems Journal in 1987. The EA structure came as a reaction to the expansion of business innovation, particularly during the 1980s, when PC frameworks were simply grabbing hold in the working environment. Organizations before long acknowledged that they would require an arrangement and long-term procedure to help the fast development of innovation and those remaining parts genuine today. Present day EA systems currently stretch out this theory to the whole business, not only IT, to guarantee the business is lined up with computerized change methodologies and mechanical development. EA helps spread out how data, business, and innovation stream together. This has turned into a need for organizations that are endeavoring to stay aware of new advancements, in areas like the cloud, IoT, AI, and other rising patterns that will provoke digital change. Good EA framework considers the most recent advancements in business forms, hierarchical structures, data frameworks, and innovations. A definitive objective of any EAP methodology is to improve the productivity, practicality, and unwavering quality of business data.

## ENTERPRISE BUSINESS ARCHITECTURE

The Enterprise Business Architecture characterizes the formal connection between the undertaking industry system and the outcomes anticipated from supporting key activities. The EBA gives a solitary source and far-reaching store of learning from which corporate activities will advance and build connections. The advancement happens from a completely incorporated venture model of the business to all IT, authoritative, and security structures. The EBA likewise gives mix capacities to programming advancement, bundled programming setup, and procedure improvement activities. Business Architecture segments incorporate business capacity maps, esteem streams, process models, frameworks and applications, information, structures and jobs. Enterprise business architecture is a multidimensional business perspective on abilities, end-to-end data, and structure, and it could also help link a business with its key partners. EBA is the extension between the plans of action and organization procedure on one side and the business users of the activities on the opposite side. It frequently empowers the Strategy to Execution philosophy. The expression "business engineering" is regularly used to mean a structural depiction of an undertaking or a specialty unit, a compositional model, or the calling itself. The key normal for the business design is that it speaks to genuine parts of a business, alongside how they connect. It is created by an interdisciplinary practice region concentrated on characterizing and examining worries of what business does, how it does it, how it is sorted out, and how it understands competition. It is utilized to plan, focus, structure, and make procedures by influencing existing qualities, and distinguishing potentials that would advance the businesses' destinations and development. Results of EBA activities are utilized to create plans, settle on business choices, and guide executions. The Zachman Framework is a prominent enterprise architecture framework highly utilized by business planners. The structure gives insights into key venture ideas that are characterized by six interrogative classifications: What, How, Where, Who, When, Why viewpoints. These questions enable the key planners who are Executive, Business Management, Architect, Engineer, Technician, and Enterprise to have the best ideas as it relates to current business situations. While the executive point of view is concerned with the extension and setting of the business, the Business Management point of view is concerned with business definition models.

## ENTERPRISE ARCHITECTURE FRAMEWORK

**The Open Group Architecture Framework (TOGAF):** This enterprise architecture model has proven to improve business efficiency. It is the most prominent and reliable Enterprise Architecture standard, ensuring consistent standards, methods, and communication among Enterprise Architecture professionals. Those professionals who are fluent in the TOGAF approach enjoy greater industry credibility, job effectiveness, and career opportunities. This approach helps practitioners avoid being locked into proprietary methods, utilize resources more efficiently and effectively, and realize a greater return on investment. TOGAF is an enterprise architecture framework structure for big business engineering that gives a way to deal with planning, arranging, executing, and data innovation design. TOGAF is displayed at four dimensions: Business, Application, Data, and Technology. TOGAF enables organizations to adjust IT objectives and large business objectives while arranging cross-departmental IT activities. TOGAF enables organizations to characterize and sort out prerequisites before a task begins, keeping the procedure moving rapidly with minimum errors. The Open Group Architectural Framework (TOGAF): TOGAF gives standards to structuring, arranging, executing, and administering organization IT design. The TOGAF system enables organizations to make an institutionalized way to deal with EA with a typical vocabulary, prescribed benchmarks, consistent techniques, proposed instruments, programming, and a strategy to best practices. The TOGAF system is generally well known as an undertaking engineer structure, and as indicated by The Open Group it's been received by in excess of 80 percent of the world's driving organizations. TOGAF enables organizations to execute programming innovation in an organized and sorted out way, with an emphasis on administration and meeting business goals. Programming advancement depends on a joint effort between numerous divisions and specialty units both inside and outside of IT, and TOGAF helps address any issues around getting key partners in the agreement. However, you don't need to embrace all aspects of TOGAF. Organizations are in an ideal situation assessing their requirements to figure out which parts of the structure to concentrate on. The Open Group created TOGAF in 1995, and in 2016, 80 % of Global 50 organizations and 60 % of Fortune 500 organizations utilized the structure. TOGAF is free for organizations to utilize inside, yet not for business purposes. Be that as it may, organizations can have devices, programming, or preparing programs confirmed by The Open Group. There are at present eight guaranteed TOGAF devices and 71 certified courses offered from 70 organizations across the world.

**The Information Technology Infrastructure Library (ITIL):** The ITIL characterizes the structure and skillset needed for data innovation and a lot of standard operational administration techniques and practices to enable the organization to deal with an IT task and related infrastructure. ITIL incorporates service strategy, service design, service transition, service operation and continual service improvement as innovation change. Strategy describes business objectives and client necessities and how to adjust destinations of the two elements. Design outlines practice for the generation of IT policies, models, and documentation. Transition advises on change management and practices, and it helps to guide administrators through environmental interferences and changes. Operation offers approach to oversee IT benefits on a day to day, month to month, and yearly premise. Service Improvement covers how to present upgrades and approach policies inside the ITIL procedure structure. ITIL is a structure intended to institutionalize the determination, arranging, delivery and maintenance of IT benefits within a business. The objective is to improve effectiveness in service delivery instead of simply back-end support.

**COBIT**: COBIT represents control objectives for information and related technology. It is a structure made by the ISACA (Information Systems Audit and Control Association) for IT administration and governance. It was intended to be a steady tool for managers to use in connecting the significant hole between technical issues, business dangers, and control. COBIT can be used in any organization no matter their industry. COBIT guarantees the quality, control, and unwavering quality of data frameworks in organizations, which is likewise the most significant part of each cutting-edge business. Today, COBIT is utilized by all IT business process chiefs to furnish them with a model to convey an incentive to the organization and improve IT-related practice. COBIT ensures that honesty is in the data framework. COBIT integrates process description, framework, control objectives, maturity models, and management guidelines in the IT process. The framework helps in sorting out the objectives of IT administration while connecting business necessities. Descriptions are a typical language for each person in the organization, and the procedure incorporates arranging, building, running, and observing of all IT processes. Objectives give a total rundown of necessities that have been considered by the administration for successful IT business control. Models access the development and the capacity of each procedure while tending to the holes. Guidelines help in allocating obligations, estimating exhibitions, conceding to basic targets, and outlining better interrelationships with each different procedure. COBIT is utilized by both government and private segment associations since it helps in expanding all IT related frameworks within a system.

**The Project Management Body of Knowledge (PMBOK):** Task Management has dependably been practiced casually over the years, yet started to rise as an unmistakable calling in the mid-twentieth century. The PMBOK portrays the whole amount of hypothesis and discovering that incorporates the task that supervisors undertake in field projects, including formally composed structures of training just as already casual and unwritten knowledge. As the PMBOK incorporates dynamic, integrative procedures that are in steady advancement as training develops and changes, the PMBOK is refreshed on a regular premise. Like different controls, the information is characterized, tried and changed by the people who use and cultivate expanding learning inside the domain of organizational leadership. PMBOK is somewhat challenging to persue since it is a depiction of the procedures. It is a production of the Project Management Institute (PMI). The first variant of the PMBOK was distributed in 1987, the second form in

1996, the third form in 2004, the fourth in 2008, and the fifth in 2013. Individuals from the PMI can download an advanced duplicate of the PMBOK for nothing. A project is usually regarded as a success when it accomplishes the targets within the time frame and budget. The primary reason PMBOK is significant is that it enables organizations to institutionalize practices in every office. PMBOK can help supervisors in organizations work with an institutionalized framework in all offices and branches of an organization. PMBOK examines what works. The strategies recorded inside the task can help individuals who are unsure of how to attempt projects. PMBOK additionally talks about what doesn't work. When organization supervisors put time into learning the principles, they likewise are putting the time in figuring out how and where standards can be broken. PMBOK takes into consideration reducing project costs organizations funds by placing a limit to unnecessary spending on a given project while considering quality control.

To sum up, developing a successful enterprise business architecture framework requires some level of understanding of the competition you have in the business, possibilities of partnership and a consideration of coopetition. What does competition, partnership and coopetition have to do with building a successful business framework. The next chapter extensively highlights these factors with respect to developing an adaptable business in this digitalized new world.

# CHAPTER 13

## COMPETITION, PARTNERSHIP, AND COOPETITION

As we have seen in the previous chapters, digital transformation can provide a competitive edge, and enable an organization to capture market transition while building an innovative culture and outperforming their competition. In this chapter, we will address three areas: Partnership, Competition and Coopetition in ecosystems, as it relates to developing a business architecture framework in a digitalized business world.

## OVERVIEW OF COMPETITION

Whenever there are numerous buyers and sellers, competition is bound to happen. It becomes survival of the fittest. To stay in competition, sellers have to strategize product offering, reduction of price, design, advertisement, proximity, customer service, user experience and any other strategy that would keep customers' patronage and loyalty. At the point when customers have numerous needs, organizations must stay on their toes and keep on offering the best high-quality products at an affordable price. Competition controls the free market activity of business sectors, making product offers very reasonable. Under a highly competitive market, no organization can monopolize price since customers have several options they can buy the same product from. Competition creates an opportunity for both old and new organizations to have their target markets. When business sectors do not have enough competition in their industry, prices tend to go higher. Competition is one of the major drivers of innovation in industries; it gives room for cooperation and demands in the market. Organizations that offer the equivalent of comparative products would eventually compete for market share and profitability. The more competitors enter into a market, the more challenging it becomes. Rivalry brings down costs as organizations vie for clients and a piece of the pie. It's significant for entrepreneurs in an organization to understand how competition effects profitability and brand positioning. The entry barrier into the market is determined by the level of completion because completion affects various aspects of every business sector. The entry barrier is low when there is a high level of competition. In competitive markets, it is harder to enter the market and contend with the existing brands; this could be because of expense or the time it would take to gain market share fully. Another way competition influences a business is in a value set. In competitive enterprises, a business should be aware of its substitute pricing strategies. Industrial markets focus on mass selling, while customer markets involve breaking the mass. Digital transformation has made it easier for organizations to go global. Organizations have gone worldwide, showcasing their products and meeting customer needs across the globe. Markets for consumers have a lot of marketing activities to sell fast moving consumer goods like television, smartphones, watches, fruits, milk, refrigerators, bags, toys, air conditioners, books, laptops and so on. As customers' marketing of products gets harder, consumer markets have the highest product substitutes, including products with high quality and imitation products, leaving customers confused in making decisions because it is difficult to differentiate the quality standards of each brand. Brand reliability is minimal in consumer markets because it is the product that is most available in the market that customers buy.

There are many diverse market structures that can describe an economy such as monopoly, oligopoly, and perfect competition. Every one of them has their very own arrangement of attributes and suppositions, which influence the basic leadership of firms and the benefits they can make. In monopolistic competition, one business serves a whole market. This business is the sole distributor of its product with no challenge or contenders. This absence of buyer decision as a rule results in high costs, as an imposing business model is made. For example, when a legislature is the sole controller of a product, like power, gas, mail, or railway, there is going to be very little innovation in the business. Pfizer could charge anything it desired for Viagra since there was no evident substitute for the medication. Today, Viagra is accessible in nonexclusive structure, dispensing with Pfizer's syndication. Monopoly could exist; as a result, a patent given to an inventor which gives full rights to the invention and shields against the competition, thereby successfully assuming control over a market. Markets must continue to be available to new competitors if costs are to remain low, and products are to meet demands continually. An oligopoly portrays a market structure which is dominated by just a few firms who have the required license to operate. The organizations can either go up against one another or work together. An oligopoly is where there are multiple contenders, however close to a bunch. For the most part, oligopoly markets have a high boundary to passage. One memorable case of this is railways. Just a couple of organizations were given the correct licenses to manufacture railways, and just a couple of organizations had the cash. In oligopolies, all organizations are in danger of entering a value war, which can, at last, be hurtful to a business' primary concern. Overall revenues will, in general, be higher in oligopolies in light of the fact that there is little challenge. Normally, governments set laws to regulate oligopolies. Perfect competition depicts a market structure, where a larger number of firms contend with one another. In this situation, no firm has any critical market control. Therefore, the industry overall creates the socially ideal dimension of yield, since none of the organizations can impact market costs. Perfect competition expands on various suspicions that all organizations augment benefits; there is free entry and exit to the market; all organizations sell identical products; there are no customer preferences. Technology has significantly diminished transaction and communication costs in vertical firms. Disruption and uncertainty have made firms adapt to changes in both customer needs and their ecosystems. Firms can influence the market when they partner with other stakeholders in the economy. New industrial structures are developing based on partnership within the ecosystem. For instance, organizations, like Uber and Bolt, depend intensely on the gig economy. Technology firms have a boundless social permit; in the following decade, they should explore issues like trust, security, regulation and privacy issues of their users because competition would get tougher, and any loophole in business model would give rise to a business opportunity for other competitors.

## BENEFITS OF PARTNERSHIP

Ecosystems have given rise to high network partnerships and collaboration in various industries. Today business ecosystems have become a fundamental piece of the development plan in an organization's lifecycle. Sales and Marketing Partners help with the distribution of products within their networks; they help drive sales. White Labelers or OEM Partners give proprietary solutions with customized branding. Technology Partners are fundamental in today's business terrain because they enable an organization to enter into new markets without a high transaction. Organizations can significantly diminish expenses when they have a strong partner network. Great business partners would always provide solutions to business issues; taking a look at the solutions technology partners, distributors, service partners, and other key business partners. Providing regularly is what keeps organizations outstanding. Cooperation is good when there is trust within partners. Traditional business needs the help of digital partners to make it through digital disruption. A partnership begins with understanding and valuing each other's contributions. Collaborating pushes away limitations, helps organizations break into new markets and empowers organizations to jump the conventional hindrances of development and scale by utilizing the core capabilities of each partner. It is necessary that every partner respect laid down rules and standards. This gives each organization the certainty that they are not compromised nor are their trade secrets revealed to their competitors.

Fintech is one industry that has greatly benefited from the partnership; the sector has innovatively disrupted financial services by creating market opportunities for all players. Fintech is basically endeavoring to make payments secure and seamless. The over-dependence on physical cash keeps several people out of the worldwide economy. It has been evaluated that in excess of 2 billion individuals are essentially underserved by the financial system framework. PayPal has been able to set up key partners across the globe, to extend financial inclusion to the unbanked. They have spread their partnership across various sectors like technology companies, retailers, government, social media, entrepreneurs, or e-commerce. A coordinated effort is key as businesses become progressively advanced. Collaboration is not only in the Fintech industry but in several industries that are trying to improve the customer experience. Working together to execute innovation to address long-standing difficulties or totally retool how we have recently tended to circumstances over the years is a pointer that every industry needs a strategic partnership to scale up in this digital era. Digital partnership can truly spark innovation. For instance, Spotify, the Swedish music streaming company made major headway as an industry pioneer, when they entered into digital partnership and transformed their value proposition. With US espresso chain Starbucks and Japanese stimulation and innovation organization Sony as strategic partners, Spotify was able to get a higher market segment of the millennials connected to their platform. Sony announced that 1.5m individuals exploited its new Spotify service on the very first day of its product launch. For digital partnerships, timing is crucial. Digitization implies that organizations must build their capacities and stay aggressive and alert to changing market conditions.

A portion of these partnerships will thrive, and others will fail. Organizations will discover it progressively harder to compete today without strategic partnerships because everything is becoming connected from smart homes to big data and emerging technologies. Digital partnerships are simpler to enter, less expensive to be associated with, and simpler to exit; doing nothing is presently the most hazardous alternative. Digital partnerships are progressively getting more popular, whereby non-digital companies are uniting with the most advanced tech businesses to create something new for the market. Organizations like Apple, IBM, and Samsung have all formed strategic partnerships to foster creativity and innovation, which has helped them serve regional and global markets faster and more effectively. Companies like Uber, Airbnb, and Netflix concentrate on what their customers need and improve their experience. Building strategic connections within an ecosystem would build up a group that is intentionally mining and dissecting the data given by the ecosystem to control events and leadership in any sector. Organizations need to adjust to and grasp digital change, or they will get left behind. Human survival relies upon our capacity to detect and react to changes in our society, and organizations are not different in any way. Businesses need to go digital and consider partnerships that would help them remain sustainable.

## CONSIDER COOPETITION

Coopetition consolidates the concept of collaboration and competition. Coopetition requires innovative reasoning, competitors can frequently profit by participating with each other deliberately. Competitors are not just people whom we should fight against for customers, region, and a piece of the pie. While organizations in an industry may be comparable, they normally serve various customers and have various qualities and shortcomings. This can leave holes between the products they offer, and the necessities of the customers. Coopetition wipes out these holes. Coopetition takes a look at a market and perceives that competitors can likewise become suppliers, as on account of Microsoft and Intel. It takes a horizontal view at the two, perceiving that an organization may have a great market entrance in one district, while other regions are almost impossible. Coopetition perceives these sorts of contrasts as open doors for organizations to complement each other, as opposed to contending for market share. This methodology change contrasts with development opportunities for the parties involved. Competitors at that point become teammates, creating joint arrangements that totally satisfy all customer's needs. Coopetition is part of the elements of human life. In nature, competition is for development; in business, the rivalry is for enduring and flourishing. In the storehouse mechanical age, the challenge is tied in with controlling to keep business as usual; and venturing into a more profound advancement. Competition is a piece of the hereditary inclination of each living thing in nature as a survival-chasing mentality. The results of the competition can be a terrible and aggressive mindset with negative outcomes. The healthy challenge is significant for either change or advancement. Coopetition changes the standards. It enables organizations to consolidate their mastery with that of others to make an exceptional product or customer. Businesses develop and advance after some time. Coopetition can enable businesses to differentiate by working with others in a related field.

Coopetition at the business makes obstructions that surface between divisions inside an organization, effectively constrained, causing individuals who should be in a similar group to look out for one another. Coopetition can improve profitability and business development. Many indigenous businesses have great reach in their home markets; however, outside that geographic territory, they are not outstanding. Coopetition encourages market entrance; by working cooperatively with associations within target markets. Two heads are superior to one; organizations looking for ways to improve and grow need to consider coopetition as an approach to expand on what others in the industry are doing. Two competitors can work cooperatively to grow a joint venture by utilizing the plans and procedures they have effectively created. The IT business relies on coopetition. Coopetition is normal in the IT industry. Regardless of whether it is channel deals, up-selling, product packaging, or mix, innovation is a factor making a greater open door for coopetition. Any organization that needs to develop or adjust rapidly to change would do well to search cautiously for chances to work with others in their industry to give an increasingly complete answer for customer needs. Coopetition is a practical procedure for staying aggressive in a quickly evolving industry. That is the reason organizations like Microsoft, and numerous other technology giants are looking towards coopetition to improve the customer experience while lessening the expense and overhead of working together.

Overall, apart from leveraging on a critical understanding of the level of competition in the market, harnessing all the benefits of partnership and implementing coopetition insights into a business framework, for a business to stand the test of digital transformation, it must be developed to be adaptable to the ever changing and transforming world. What better way is there to be adaptable to the ever-changing world than to build an innovative culture that gives room for any level of disruption that comes with technological and industrial revolution? We will be discussing all about building an innovative culture in the next chapter.

# CHAPTER 14

## BUILDING THE INNOVATION CULTURE

Culture will eat strategy for a snack, as Innovation seems to accelerate the pace of change and Design Thinking Methodology in every industry. This chapter will provide details and examples on the importance of Innovation.

Ideas are a basic piece of the innovation culture. Indeed, creativity cannot flourish without imaginative critical thinking. In a genuine innovative culture, each worker's thought is esteemed and given equal importance; nobody is denounced or disregarded when they offer a potential critical thinking proposal. Rather, they are commended for their commitment and offered the chance to work with the idea, so as to check whether it can transform into something bigger. Innovative culture in an organization is an ideal approach to use the current ability inside an organization. Everyone becomes excited to offer their skills and abilities to the organization since they realize that their thoughts will be esteemed and that the idea will be executed when it works. Working environments that encourage a culture of innovation, for the most part, buy into the conviction that advancement can come from anyone's idea even if they are not at the management level. Innovation is all-encompassing; every industry needs to genuinely develop all through their organization, by creating a culture that energizes participation, rewards innovativeness, and cultivates a positive working style that makes more open doors for each person. Everyone in an organization should be encouraged to bring up ideas. To promote innovation culture, individuals need to realize that they matter, their thoughts matter and that their commitments will be a positive expansion to the organization in general. Numerous workers are crippled by the inclination that they are unfit to settle on choices without running them by another person first. The entryways for positive communication must first be opened before an organization can experience collaborative innovation. Success comes most times by risk-taking; therefore, people need to leave their comfort zones so they can think creatively outside the box. Execution is basic, at the point when thoughts are essentially examined, nothing changes. It might be increasingly useful to have a whole group teaming up to tackle an issue.

An innovative culture enables individuals to explore the world. Discovery is linked to innovation. Google has a 20% time they give to their employees to work on something they love, one-fifth of their work hours is dedicated to fostering an individual innovative culture, and this method allows staff to be highly creative and smarter. Innovation culture is hard to build up, and support yet is considered very fundamental for making headway in a highly competitive world. With regards to IT innovation, it is significant that IT professionals understand how to monetize innovation. Sustainable innovation culture can be achieved by embracing digital transformation, establishing innovation labs, and rewarding discovery. Innovation is not bad, but at the same time not enough to move from conventional to digitalized business forms. Today digital innovations are firmly attached to business development. In most organizations today, there are certain metrics put in place to measure individual contribution and level of innovation. The success of innovation can be accessed by value-added and sustainability of innovation culture. Digital transformation has changed almost all phases of human endeavor. When we measure innovation from communication, transportation, healthcare, governance and other areas, we can see a lot of value has been added from the first industrial revolution to the fourth industrial revolution.

## CASE STUDY ON INNOVATION CULTURE

In 1955, Fortune Magazine recorded the 500 biggest organizations in a rundown that were synonymous with progress. After 60 years, just 71 of those organizations remain, while the other 729 companies are nowhere to be found because they failed to innovate. The 729 organizations that neglected to adjust to changing business needs missed seeing a wave that made their reality unimportant. Kodak and Walkman ought to be liable for neglecting to develop and missing the waves that cleared their industry. Nokia once led the mobile industry market when Samsung was battling, and Apple was still not part of the market. Nokia neglected to envision, comprehend, or arrange itself to manage the evolving times. In spite of the fact that they made a return with Windows Phone after the Microsoft procurement, they are as yet attempting to return to their previous position of being the market leader.

There are many such contextual investigations of disappointments because of the absence of innovation culture. Myspace failed to understand the importance of user experience; the loophole made Facebook take advantage of where Myspace failed to proffer innovative solutions. Myspace core business was centered on connecting people within a single space whereas that of Facebook was on fostering connection across the globe. In the end, Myspace lost to Facebook, which made sense of this business gap, that people needed to interface on different dimensions, and this learning transformed Facebook into a worldwide power.

Blackberry concentrated more on business activities within their organization as opposed to customer needs, while Apple and Android though not the market leaders then, were able to fully capture consumer needs and perception as their core business activity. Apple concentrated on touchscreen smartphones, while Android was an open system platform that was affordable. The offers from Apple and Android were very astonishing to a larger number of the world population; this was how they became the market leaders of the mobile industry. Apple re-imagined the market and left Blackberry, who was blinded by their very own initial achievement, faltering.

The failure of Boarders to maximize e-commerce made them lose their market share to Amazon, who took the opportunity of the internet to sell all products and also deliver products globally. Failure to innovate into digital technologies made the once-powerful Borders book shop redistribute its site to Amazon.com, Inc.

Apple, at some time, needed $150 million to remain in business. Due to its innovative culture, in a brief timeframe, Apple became one of the most significant organizations ever. Disruptive innovation made Google market leader in the area of search engine and Ad selling. It was a start-up trend-setter which changed into a valuable global company. Everything in Google experiences a design thinking process and pursues high innovation standards. They likewise approach extraordinary innovations and buy smaller innovative companies. Innovation centers around making a difference by forecasting current and future trends in markets, technology, consumers, demographics, population, fashion, perceptions, preferences and adjusting to changing innovations to improve on business processes, services, people and profitability to gain sustainable competitive advantage.

## CREATING AN INNOVATION CULTURE

The way of life of an organization decides the connections and activities of its internal and external environment. A positive innovation culture makes motivations for workers and prompts an expansion in the imaginative quality of the organization. Culture depicts the level of development that would take place in an organization. Innovation culture works as a sort of cross-cutting society, whose models and qualities are molded and strengthened by all who take part in the process. The absence of information on innovation strategies frequently reduces the imaginative capacity of people.

Organizations need to promote cross-departmental and cross-divisional development workshops with balance by an outside specialist, inner learning from recently gained or effectively existing information from instructional classes, utilization of new procedures and innovations. There has to be an eagerness to advance; creativity is never an act of request; it is not ordered for. It must be naturally propelled so as to build up an enduring innovative culture in an organization. There has to be risk-taking when trying to build an innovation culture; workers can commit errors and gain from them and will always be motivated to bring in their best. Innovation can happen on a huge or small scale. It tends to be as basic as making a procedure simpler, setting aside some cash in our savings, buying a house, shopping, or figuring out how to make tasks more productive. An innovative culture is the work atmosphere that top-level management develops to support design thinking and creativity in their organization. Sticky notes or whiteboards can help an organization consciously practice sustainable innovation culture; they come in handy as the best innovative tools during brainstorming sessions.

Digitization depicts the procedure by which an organization frames a system to actualize innovation to improve business and fulfill the regularly changing needs of the customer. It is significant for present-day organizations to perceive innovation's potential and acknowledge when the ideal opportunity for change has come. At the point when a new product enters into the market, high demand can occur; likewise, a whole industry can change instantly as contenders race to make something better. Purchasers benefit the most, in light of the fact that this disruption creates a rivalry, which prompts better product offering in existing and new industries. The rivalry has expanded territorial, national, or worldwide because of more extensive access to innovation and information sharing through online and offline platforms. As indicated by Sylvie Laforet, the variables driving development found by past scientists are the market condition, organization development, CEO's drive, rivalry, customers, and disruptive innovation. The results shows that corporate policies, emerging technologies, rivalry in the market, new ideas, customer and market demand, those along with others are significant innovation drivers.

Building an innovation culture helps organizations stay relevant, stay competitive, address difficulties and remain focused. One approach to profit from innovation is to commercialize the idea, it must be thoroughly researched to make sure there is a demand, because every innovation requires sustainable growth and profitability. Thereby, protecting innovation is fundamental. Innovation is essential to the success of any organization. It makes fresher, quicker, and better products for customers, and furthermore enhances innovation culture within an organization. It has become basic for organizations to develop with innovation as a focal point. The individuals who won't change and develop in the feeling of innovation and mentality will battle to remain pertinent in a competitive market place.

Enthusiasm is highly needed in building innovative culture; the capacity to advance comes with confidence and high spirits from everyone involved in the innovation process. Capacity will be made that offers workers the chance to think and act creatively and innovatively. Clear objectives, duties, and procedures with respect to idea generation and the current issue must be put into consideration. Cross-divisional innovative groups can be put in place when building an innovative culture. Innovation does not necessarily have to be progressive, new, or improved; innovation has to be sustainable and continuous. While Dell enhanced frameworks and procedures, Toyota adopted a steady strategy for development. Innovation can be in advertising, new products, existing products, processes, procedures, branding, technology, management, product offerings, and other core business areas. Today, business mantra is about innovation to survive competition, thereby innovate or die can be seen by most big organizations which we would portray below.

## INNOVATION AND DESIGN THINKING

Design Thinking helps an organization develop new ways of thinking, such as intellectual and practical ways. Design Thinking could be introduced into new products, machines, proposals, or buildings whereby designers have to come up with better approaches for deduction so as to configure better solutions that take care of our present issues. Design Thinking is connected with innovations in product offerings within various industries. Design Thinking is a component of creativity. It is the capacity to take abstract, unique ideas and develop a product from that vague idea. Like the concept of the digital twin that was introduced in an earlier chapter of this book, it is a fact to say that most digital technologies have a fundamental root in both Design Thinking and innovation.

Disruption in human development since the industrial revolution has made several researchers think of ways to strategize on every business and economy to the point of getting the most profit by the minimal measure of time and assets. While this may have had some level of accomplishment on the dimension of profitability and effectiveness, Design Thinking focuses on strong, new human-focused methodologies, thereby, drastically changing how we approach, investigate issues, and discover answers for those issues, helping us transform from traditional models into digital transformation. Other than the progressing battles between the analytical and creative sphere, different elements have significantly changed the manner in which we see, get, understand, and translate our general surroundings. Technology keeps developing at such a fast pace that rapidly changes business operations.

Design thinking became so well known on the grounds that it limits the vulnerability and danger of innovation by drawing in users through a progression of models to learn, test and refine ideas. Design thinkers depend on user bits of knowledge picked up from genuine examinations, tests and activities, not simply broad information or statistical surveying that was done by another person. This methodology unites what is attractive from a human perspective with what is technologically accepted and financially suitable. Design Thinking is most appropriate when tending to issues where numerous circles diverge, at the convergence of business and society, rationale and feeling, judicious and innovative, human needs and financial requests and among other frameworks, people within a given ecosystem. Design Thinking is ideal for complex issues that affect a larger population.

## DESIGN THINKING
### A FRAMEWORK FOR INNOVATION

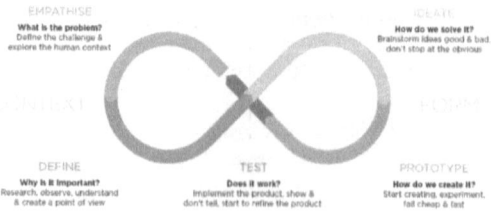

With all we have been discussing and as highlighted in chapter 4 above about navigating the new complex world, it is evident that technological revolution and digital transformation disruption is bringing about a new world. In this new world 'Data is King' unlike the popular saying that refers to cash as king. With the right use of data, you are almost unlimited with the level of creativity and innovation that anybody can bring to the table. How do you cope in a new world where data is the new currency? Find out all about how data is transforming business and effecting digital disruption in the world.

# CHAPTER 15

## THE NEW WORLD, WHERE DATA IS THE CURRENCY

In this new advanced world, Data is the world's most important resource, yet controlled and abused by a couple of huge players. Everything is information. Indeed, even you and I, are a collection of information or the bank of memory known as DNA. All that we do and communicate with as individuals are filled with information. Regardless of whether that implies setting off to the specialist, shopping, having a feast, or essentially sending an email, these are pieces of data that recount a story of our identity and what our inclinations are. Data is the reason Amazon realizes which books you like to read before you do. Data alone is characteristically feeble; it doesn't really do anything except that it is utilized. Google, Facebook, Twitter, Instagram, WhatsApp and so forth offer applications that are free of charge yet complimentary. These organizations have been estimated to be worth billions. However, there is no service fee to subscribe and creating an account on their platforms has made customers lose their privacy. Clients' information has become big data which can be sold to marketing companies and other third parties. Data is the main genuine money of the web. 90% of the information on the planet today was made in the last two years. Between now and 2020, the worldwide volume of advanced information is relied upon to increase multiple times or more. A lot of that new data will comprise of where individuals have been, what items they have purchased, what motion pictures they like and which philosophies or ideologies they support. Government is one of the greatest makers of data. Organizations are trying to take advantage of the market for individual information. Government is likewise a significant player in the information economy, as a controller as well as a noteworthy supplier and shopper of information. Most data from government establishments are accessible at zero cost. Government is additionally a significant player in the information economy, as a controller as well as a noteworthy supplier and buyer of information. Data is the new resource that enables organizations, government, and people to make smarter decisions easily and in the most effective ways possible. Figure below, from the National Human Genome Research Institute shows the exponential decline in Genome sequencing cost.

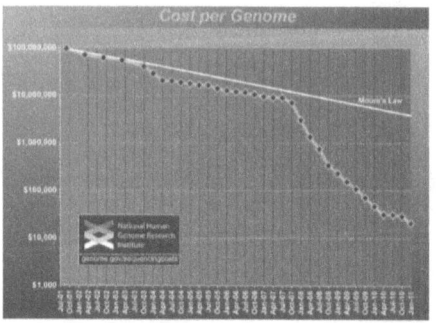

The very base of artificial intelligence and digital transformation is data. The serious nature of data computing has a number of effects on the industries involved. Mastery of data can prove to be a tool for an organization or company to use for better production. Data can be used for many effects, these include:

Creating new revenue streams — As already known, data includes a lot of information outlets and codes that could be analyzed together and used for the generation of revenue for further industrial purposes. The understanding of who needs what and how they want it served is the essence of data computation, and once these questions are solved, the companies can effectively use the information to produce on demand goods.

Reduction in cost of operations — the readily available data can help the company to cut costs of operations. For example, demographic surveys can be done on media outlets without actual physical involvement of the company's representatives. Data also reduces the risks of trial and error which might result in the wastage of resources.

Enhancing customer experience — Data computing and utilization of customers' ideas can go a long way to brand the company as being sensitive to customers and also improve the relationship between the producers and the end-users.

Enhancing security & privacy — The company can improve upon their security and privacy measures feeding directly from available data. The data can help to easily trace malfunctions, rework vulnerabilities, respond to the demands of the customers with time relativity and also improve the overall outlook of product and services.

## THE COMMERCIAL CENTER FOR DATA

**Open providers**: Government offices have gathered an enormous store of genuine data over the span of working together with all parties in the ecosystem of governance.

The United States government has been discharging huge government informational collections to the general population, for nothing. Organizations and people utilize this information to make important decisions, doing it quicker and more economically than the government could without anyone else. Through its White House Open Data Initiatives, the government has become the largest provider of open source data that has helped in both primary and secondary data of research analysis.

**Information aggregators**: Some organizations today, especially marketing companies, have been able to gather tremendous databases of consumer inclinations and practices. These companies know almost everything about their customers, from their emails to their conversations, thereby making accurate predictions on buyer wants and future needs. Joining data from open records and shopper exchanges gathered from internet-based life, versatile transmissions, and different sources, these aggregators give sponsors new bits of knowledge into target gatherings of people.

**Data for service:** It is right to say that social media platforms do not offer subscribers free service. From the previous chapter of this book we discussed that people lose their privacy once they sign up on social network platforms like Facebook, Twitter, or Google. We pay for the benefit by unveiling individual data, which is sold to advertising companies and other third parties who want a particular type of dataset.

**Information defenders:** To help address concerns identified with security and individual information, the market presently offers items to give people authority over their own data. With information storage from Personal.com, for instance, you can store individual data, control access to that information and trade it as indicated by your desires. Reputation.com also discloses to you what data others are gathering about you, who is gathering it and how they are utilizing it. A few firms, likewise, give refined protection services to keep individual information unknown and safe from the prying eyes of cyber bullies and other third parties who love to monitor people.

## PEOPLE VERSUS MACHINES

Humans are highly intelligent due to their ability to control unstructured data. Humans can translate and comprehend data without learning or having experience in data related fields. On the other hand, machines tend to assimilate structured or machine-readable data more, as compared to the ones carried out by humans, which is unstructured data. Man and machine require a lot of directions for controlling information in order to apply them in a given project or programming. Altogether for a program to perform effectively based on current information, there has to be some sort of uniform structure. Machine learning is helping devices act like humans whenever they face unstructured data. Such machines use likelihood just as past execution and ongoing learnings to perceive designs in information and make expectations. The criticism circle of the information telling the machines whether their forecasts are valid or false will, at that point, sway their future choices, in this way making it a procedure of learning. The quick headway in AI-related advances is being coordinated by ideas like compassion and passionate knowledge and the capacity to perceive sentiments both in oneself and in others. However, feelings remain a profoundly human trademark, probably not going to be done better by machines no matter how advanced they may seem to be. Emotions and collaborations will keep on requiring real people. Machines are showing signs of improvement at learning and applying their learnings to new circumstances, yet they still have a long way to go, they can never have similar creativity to an individual.

## SIGNIFICANCE OF DATA

Information is at the center of almost every business choice made. HR chiefs are gathering data from online assets to deciding the best individuals to select and affirm insights concerning them. Advertising and marketing companies are using data to segment their market, discover customers who are prepared to purchase, and accelerate the sales process and ease of doing business. By utilizing information adequately, an organization can streamline the way toward getting an item made and placing it in the hands of the client. An organization needs to create techniques for showcasing, deals, HR and tasks. Getting the correct data implies recognizing what data is critical to the organization's basic leadership process. The system begins with fundamental statistic information. At that point, it thinks about valuing dependent on the instruction and pay of purchasers, and how that bunch talks. Training and salary are significant, on the grounds that the more taught and the higher the pay an objective gathering has, the more probable it is that the business can sensibly pitch a higher-finished result to the gathering who can comprehend, acknowledge and manage the cost of the item.

As advanced communication become omnipresent, data will be in our current reality where almost everybody and everything is connected in both the virtual and real world. The fourth industrial revolution will change the way people think. The quick rate of innovative change and commercialization in utilizing digital information is undermining certainty and trust. Worries about the abuse of digital information keep on developing. In this day and age, where advanced change influences each industry, it is crucial that individuals believe that their information is by and large satisfactorily taken care of and ensured. In the event that one works with information, one must be responsible for information depended at a worldwide dimension. Digital businesses must understand the change and go for a higher standard of security. Government officials and business pioneers need to deal with finishing up a lot of global principles that permit the exchange and capacity of information while engaging individuals to control how their information is utilized.

## DATA ECONOMY

Before the coming of the age of digital transformation, most organizations' resources comprised majorly of property, plants, hardware, intellectual property, stock and money. In today's digital world, it is a different ball game as data is the new resource for organizations. Data is rising in huge volumes, and organizations keep gathering and collecting necessary information that would help them stay on top through the digital revolution. The data economy is very vast and full of lead generation data; for instance, data has made it easier for businesses to reach a larger target audience based on their similarities and perceptions. Information begins from numerous sources that are simpler to gather and investigate. Because of the accessibility and new information, more data is being traded inside and among organizations. This has produced another economy based after utilizing information to create an incentive through both inner and outer methods. IoT gadgets are one of the main drivers behind the move to this new economy. As the quantity of 'things' becomes more instrumented, interconnected, and savvy, the data they produce will develop exponentially. New methodologies and plans of action will be required. In spite of the extraordinary steps we have seen in IoT innovation and applications, the Internet of Things has much space for development. Data can enable us to find out about complex IoT frameworks. There are numerous open doors for cost regulation with IoT data. These open doors are driving IoT speculations by organizations and expanding the number of new players. As the volume expands, the information economy will proceed to develop and empower organizations through selling and trading data. As IoT empowered devices are received, access to data from those items is decreasing the cost of possession. As ecosystems rise, coordinated effort stays fundamental. With the rise of the data economy, organizations can create adaptable supply chains, and opportunities from the response of real-time consumers, consumption data and location information.

# THE ROLE OF BIG DATA

The data economy has increased the utilization of big data and analytics by big corporations. For instance, big data has helped Fintech companies give customized services by breaking down huge measures of information. The best companies have been able to improve customer relationship, language, and culture because the intensity of Big Data is based on utilizing psychological innovation. When organizations utilize AI and Big Data, they can easily customize offers that are better for their customers, and also stay ahead of the competition in the market and take the necessary steps to ensure continuous growth. Rivalry over all enterprises is just getting progressively extreme, particularly as more organizations enter conventional markets. Innovation has never been increasingly open, adjustable or moderate. It's making everything fair. Those organizations that comprehend the requirements of their customers quicker, and can convey less expensive, will win. The digitization of data leaves a computerized impression. At whatever point, we get to an application, make an installment with a bank card, read the news on the web, go buy a surveillance camera or purchase a ticket for open transportation, we make information that shows what we have done, where and when. This implies in the advanced world, there is an undeniably comprehensive record of what happens in the physical world. When there is a huge amount of data, in a perfect world, all things being equal machines will become faster and smarter at processing results.

Data is progressively developing on our identity, who we know, where we are, the places we have been to and where we intend to go. Big data being gathered is huge: from individual profiles like Facebook, Instagram, statistical information, financial balances, medical records or work profiles. Our web inquiries and destinations visited, including our preferences and purchases, food consumption and home temperatures, regardless of whether our lights are on or off. The rundown keeps on developing. Mining and breaking down this information gives us a chance to comprehend and foresee how individuals carry on at social events, gatherings and worldwide dimensions. Digital information is as significant for economies and social orders as it is full of inquiries concerning security. Firms gather and utilize this information so as to monetize it by fitting their services in line with the information gathered. Also, governments use big data information to enact policies that would deliver sustainable growth for their country. Scientists use big data to speed up the process of a scientific breakthrough in their research.

## Data Economy Giving Rise to The Gig Economy

As data rules the world in this current revolution, business frameworks built on the backbone of data across sectors and with critical innovative technology are developing multimillion dollar organizations over a reasonable short period of time. This is giving rise to the emergence of organizations known as Unicorn companies. These companies are transforming the way we do just anything, making life easier and leveraging on their abilities to provide innovative solutions to maximize revenue generation. In the next chapter, we will be discussing Unicorn companies and how the gig economy is transforming the way we work and live.

# CHAPTER 16

## THE RISE OF THE UNICORN COMPANIES

Thanks to the digital revolution, companies are emerging and horning revenues in billions of dollars without necessarily having a conventional business architecture framework. Unicorns are the leading front of start-up organizations and are privately held start-up companies valued at over $1 billion. The term was coined in 2013 by venture capitalist Aileen Lee, choosing the mythical animal to represent the statistical rarity of such successful ventures. A decacorn is a word used for those companies over $10 billion, while hectocorn is the appropriate term for such a company valued over $100 billion. According to TechCrunch, there were 279 unicorns as of March 2018. The largest unicorns included Ant Financial, DiDi, Airbnb, Stripe and Palantir Technologies. Lyft is the most recent decacorn that turned into a public company on March 29, 2019.

Technology start-ups worth $1 billion like those above, once as rare as unicorns, are now plentiful enough and old enough that there is a whole new crop of them and it is one that looks very different. Silicon Valley's current crop of highly valued tech start-ups, which includes now-household names such as Uber and Airbnb, all benefited from the spread of smartphones and cheap cloud computing. Many of these companies managed to build global empires by simply taking existing businesses such as taxis, food delivery and hotels and making them mobile.

In other words, they digitized these businesses and sectors. Some of the start-ups became giants, like Uber for instance, which is likely to reach a $120 billion valuation by this year. But as those companies have matured and prepare to go public, the opportunities and chances of disrupting old-line industries are disappearing fast. Many of the up-and-coming start-ups that may become the next unicorns now have names like Benchling and Blend, and they usually focus on software for specific industries such as the farming, banking and life sciences sectors. This is all from an analysis by CB Insights, a firm that keeps track of venture capital and start-ups. CB Insights used a variety of data, including financial health and the strength and size of the market that a company caters to, to pinpoint up to 50 start-ups that might be on the path to achieving a $1 billion valuation.

Jason Green who is an investor at Emergence, a venture capital firm that invests in cloud software companies has remarked that while software start-ups may seem boring, a lot of them are growing quite rapidly because industries such as agriculture are starting to require more software tools as they adapt to the shift in technology era.

A successful agriculture-based company like Farmers Business Network would not have been possible 10 years ago, before the proliferation of cloud computing and the "digitisation" of farming processes. Now, farms produce a lot of data, which Farmers Business Network is helping them to process and use to make decisions. "Agriculture is going through a digital revolution," says Charles Baron, a former Google programme manager. The company charges farmers $700 a year to share and analyse data about their farms and buy supplies. It also helps them sell crops. Baron says the start-up has at least 7,700 farms as customers and has managed to raise nearly $200 million in funding.

# THE GIG ECONOMY

The Gig Economy is fast growing as freelancers and professionals are taking advantage of its service offer due to the disruption in the existing business model and technology innovation to give superior support to their clients. The Gig Economy is otherwise called the easy entry economy; it has become one of the easiest approaches to request for short based services. We have all caught wind of Airbnb, Uber and the App Economy (what Slack, as an effective business has a couple of things to encourage each and every business out there), the eventual fate of work, the fate of the working environment and what many start-up founders make progress toward is to take their business to "Unicorn" status. You may ask what a "unicorn" has to do with business. This is only an allegory to portray a business which is secretly held and has a valuation in an overabundance of $1 billion, but that has already been explained in the section above.

The gig economy provides brief, adaptable, on-the-go jobs for freelancers, professionals, and organizations looking toward hiring part-time staff as opposed to full-time employment. A gig economy undermines the conventional economy of full-time career employment. Times have changed. A lifetime profession is no longer the dream for millennials, they prefer a job that is flexible and less rigorous. The rise of the gig economy has made transitory jobs and low entry jobs available for anyone seeking such options. The gig economy is less expensive, with progressively effective services, for example, Uber or Airbnb, for those ready to utilize them. The individuals who do not take part in utilizing innovative services, for example, the Internet, will, in general, be abandoned by the advantages of the gig economy. Urban communities, in general, are the most settled in the gig economy. While not all businesses incline toward procuring contracted workers, the gig economy pattern regularly makes it harder for full-time staff to grow completely in their career since transitory workers are frequently less expensive to contract and are progressively adaptable in their accessibility.

America is well on its approach to building up a gig economy, as 33% of the working populace is, as of now, in some gig platform. In this digital era, people prefer to work remotely or from home in order for them to maximize their skills. For instance, some major cities that have traffic congestions in the morning when workers are on their way to work, the lost time spent in traffic is not advantageous to any business. This encourages autonomous contracting fills in the same number of those employments that don't require the specialist to come into the workplace to work. Businesses likewise have a more extensive scope of candidates to browse as they don't need to enlist somebody dependent on their nearness. The openness and trust the gig platforms have enabled in their business models has made it possible for freelancers to get paid without much stress. A gig economy is a free market framework wherein brief positions are normal and employers' contract with freelancers to carry out a certain task. The expression "gig" is a slang word signifying "work for a predetermined timeframe" and is commonly associated with artists. However, in the workforce, the gig economy incorporates specialists, and self-employed professionals to enlist on a certain platform based on their niche.

The gig economy or independent economy varies from conventional work in that occupations are not changeless, yet more explicitly, the term identifies with numerous irregular undertakings or individual move assignments. Notwithstanding, the term may likewise be utilized to reference longer-term independent courses of action and free contracting assignments. The pattern towards a gig economy started as an investigation by Intuit. Intuit anticipated that by 2020, 40 % of American laborers would be self-employed individuals. In the digital age, the workforce is progressively versatile and work should progressively be possible from anyplace, with the goal that activity and area are decoupled. That implies that consultants can choose among impermanent occupations and undertakings around the globe, while bosses can choose the best people for certain activities from a bigger pool than that accessible in some random zone. Digitization has likewise contributed to a reduction in office jobs as programming replaces a few sorts of work to boost time proficiency. Different impacts incorporate budgetary weights on organizations prompting an adaptable workforce and the passageway of the Millennials' age into the work system. The present truth is that individuals will, in general, change occupations a few times all through their working lives, and the gig economy can be viewed as a development of that pattern. In a gig economy, organizations spare assets regarding benefits, office space and preparing. They likewise can contract with specialists for certain undertakings that may be too extravagant to keep up on staff. From the point of view of the consultant, a gig economy can improve work-life balance over what is conceivable in many occupations. In a perfect world, the model is fueled by free workers choosing occupations that they are keen on, as opposed to one in which individuals are constrained into a position where they are unfit to achieve work. The gig economy is a piece of a moving social and business condition that likewise incorporates the sharing economy, the vibrant economy and the bargain economy. Outsourcing has been a thing for quite a long time. However as of late, the quantity of self-employed people has soared. Millennials place a higher incentive on flexible work. This is one of the reasons the gig economy is setting down deep roots. The expression "gig economy" alludes to a general workforce condition in which commitment, impermanent contracts, and free contracting are typical.

## HOW THE GIG ECONOMY WORKS

Independently, a gig, an individual errand, task, or occupation speaks to a little bit of a specialist pay. At the point when specialists total their pay from various customers or organizations, their combined profit can be like that of all-day business. The gig economy works on innovation stages that mean to interface specialists searching for adaptable work plans with the organizations who need them in a brought together area, for example, an application or site. A few stages are centered on specific specialties, for example, driving, graphic design, fashion, groceries, or distribution, while others are more extensive, interfacing gig specialists with organizations and customers for assignments extending from housekeeping, security, catering or programming. In the office, setting staff work based on their job descriptions, which means their customers pay them a settled upon rate for services rendered. In a gig economy, specialists are in charge of their charges based on the contract offer they give to their clients, they also get the opportunity to exploit the tax cuts of working their own business.

People from the workforce, with full-time employment, who need to enhance their pay, can undoubtedly get a couple of gigs in the nights or on the weekends. Gifted experts can apply more authority over their profession direction by taking part in the opportunities of some gig platforms. Free workforce benefits specialists as well as organizations who can procure the cost reserve funds of enrolling outsource assistance, without the managerial expenses of hiring full-time staff. The expense of enlisting help is frequently less contrasted with the expense of contracting a full-time staff. Organizations can, without much of a stretch, influence talented experts for certain tasks, who might some way or another be unreasonably expensive for a developing organization to keep up as a full-time staff member. With digitization undermining some conventional occupations, the independent economy can give professional stability, however not in the traditional sense of job security in the office setting. The work plans in a gig economy are frequently found through freelance networks, direct contact with connections or through advertising. Gig economy gives specialists the choice to a hyper-adaptable method for work. The gig economy has been on the expansion in numerous nations in the previous couple of years.

## IS THE GIG ECONOMY ON THE RISE?

The present pattern in the activity market is progressively pointing towards the gig economy. As indicated by the American Action Forum, the quantity of gig economy members expanded from 8.8% in 2002 to 14.4% in 2014. This demonstrates the segment is developing quickly as the general work over a similar period expanded uniquely by 7.2%. Despite the fact that there's little understanding, regardless of whether the gig economy will surpass the conventional activity, the pointers show premium and support have expanded, frequently quicker than traditional work showcases. In the UK, just 8.7% of the workforce was independently employed in 1975; however, by 2008, the number had ascended to around 12%. The Telegraph called attention to how the number in 2016 had risen further to 16%.

Digitization of the work environment has had a noteworthy impact in expanding enthusiasm towards independent work because work is becoming increasingly adaptable. The innovative transformation has had an effect by directly outsourcing all kinds of work with programming and the reduction of the time it takes to play out specific tasks. The gig economy permits workers to perform errands from anyplace on the planet. Since the errands have been digitized. The Millennials have additionally livened up the gig economy. Twenty to thirty-year-olds – the age with birth years extending from the mid-1980s to 2000 – are relied upon to make up around 75% of the worldwide workforce by 2025. Millennials are as of now exhibiting a much higher turnover rate than different ages. Millennial Branding report from 2013 featured how Millennials, once in a while, remain in a similar activity for over three years. They likewise appeared higher in working for themselves. Their goal is to seize opportunities and enjoy a work-life balance, just as working for themselves implies they could be increasingly headed to independent work. Besides, an investigation led by Elance (the outsourcing stage is presently known as Upwork) appeared to have over 80% of Millennials seeking independent work or outsourcing as a foundation of their profession procedure. The Wall

Street Journal composed an article in 2015, discrediting the thought that the gig economy is making an upset in the working environment. The article took a look at measurements in the US and asserted that as opposed to transforming into a country of consultants, "Americans are ending up marginally more averse to act naturally utilized, and less inclined to hold numerous employments". Independent work has dropped in the US from 8.5% in the mid-1990s to 6.5% in 2015, referring to information from the Labor Department. Moreover, the information featured a decrease in workers holding various jobs in 1995. 6.3% of Americans held more than one employment. However in 2015, the extent had dropped to 4.8%.

## IMPACT OF THE GIG ECONOMY ON ORGANIZATIONS

Organizations can profit greatly from contracting gig workers. Rather than enlisting a full-time worker for a position that is brief in nature, they can get the undertakings and spare the assets. Employing gig workers decreases the need to spend assets on things like training and office space. Although some certain gig work can likewise mean conflicting salary, less security, more expenses such as gas, utilities, or office supplies. Organizations further profit by the capacity to choose from a pool of best-skilled experts for certain projects. It's not important to prepare in-house staff for another venture, as the organization can basically get the activity to an expert. This can guarantee the most capable and talented individuals are dependably accountable for certain projects.

## HOW THE GIG ECONOMY AFFECTS FREELANCERS

Millennials are leaning towards these jobs because they seek work-life balance. Gig specialists are ready to pick the occupations that appear the most fascinating. In a full-time position, there isn't much adaptability in tolerating the work given to your work area every morning. Be that as it may, as a gig specialist, they somewhat take opportunities to secure the positions they are energetic about and which are fascinating to them and financially reward their efforts. Intuit released convincing information that the on-request economy is becoming a rising tide for freelancers and 1099 self-employed contractors are attracted to developing open doors secured in creating technology, as this would dramatically increase in size by 2020, a bounce from 3.2 million on-request workers to a workforce of 7.6 million in number. The on-request services go from bookkeeping to taxi rides to home cleaning; however, the shared factor is a workforce of workers that work when and how they want.

Understanding the ramifications of the gig economy is basic for workers, managers, and entrepreneurs to settle on choices, explore potential traps and discover accomplishments in a quickly advancing business scene. As Fabio Rosati, CEO of Upwork expressed, "The 53 million Americans who are outsourcing as of now contribute more than $700 billion to our national economy and help U.S. organizations contend and discover the skillset that is needed. This is only the beginning because the time we live in is freeing our workforce. The boundaries to being an independent expert or looking for some kind of employment, working together with customers and getting paid on the schedule are leaving fast, diminishing as a result of digital disruption in the workforce.

## GIG ECONOMY EDUCATION

Regardless of whether you need to improve your insight regarding a matter or gain proficiency with a totally new skill, there is no deficiency of online courses to help you on your way. Truth be told, there are such huge numbers of choices, it tends to be hard to make sense of which stage suits you best! When you need help updating your abilities, or that of your staff, it tends to be difficult in choosing the most effective learning style that best suits your schedule. To enable you to explore the quickly growing universe of online training, here are probably the most prominent alternatives to be proficient in any skill you desire to upgrade. Companies like Cisco have over six million students that have participated in the Cisco Networking Academy program since 1997. Regardless of their socio-economic background or gender, Networking Academy students develop the expertise to master, succeed, and lead in the digital economy. Together with our 9,500-plus educational partners in more than 170 countries, we are building intellectual capital in networking, security, and IOT technologies – the critical technical skills required by nearly every business on the planet.

**Coursera:** Coursera has collaborated with leading institutions in the United States of America and other countries around the globe to give online courses in several subjects. As of late, they have brought in innovation to their platform known as "specializations" which cover 10 distinctive course pathways, that makes it possible for students who enrolled for some online courses, and get their certificates from partner universities. Coursera has a wide range of an assorted variety of subjects accessible for anyone who signs up on their platform to choose from. Courses can be taken in any field from Computer Engineering to Hospitality Management or even a native language like Spanish. As Coursera prides itself on being available to everybody, huge numbers of the courses are either free or affordable at discounted rates. Certification is certain after course completion.

**Lynda.com:** This platform has been around for a while, they specialize in video tutorials for visual learners who can access their video library offer at a $25 monthly subscription. Lynda.com can be likened to an educational form of Netflix, while one centers on movies and entertainment the other focuses on learning and skill acquisition. Lynda.com gives boundless access to more than 80,000 videos on a wide scope of various subjects.

**Udemy:** This platform regularly introduces about 800 new courses added to their collection every month, Udemy is costlier than its rivals. Expenses shift extensively, going from $10 to $500 for various courses; the most prevalent Udemy courses in technology and business will, in general, be upwards of $100. Student reviews are the best way to choose courses on these platforms.

**Udacity:** This platform is more of a segmented target reach with focus on more of technology courses such as data science, machine learning, artificial intelligence and other related courses in technology and innovation, with a little focus on other areas. In case you're hoping to break into data science (called the "hottest career of the 21st century"), Udacity's program has an amazing list of educators from organizations like Salesforce and Facebook. Udacity's estimating structure enables you to pay on a monthly basis. Whenever you choose to opt out of a program before finishing it, you pay for the course up to that point, as opposed to complete payment.

**Khan Academy:** This platform business model is designed to act as a not for profit making educational based venture, with free access to the library, and lecturers who desire to run with their philosophy. The focus is more on traditional models of education in certain subject areas. Khan Academy gives a blend of video and content-based materials in math, economics, science, computer programming and humanities. Since Khan Academy is highly accessible for free, it is a great place to learn the basics of a subject area before moving onto a further developed course somewhere else.

**Codecademy:** Previously sponsored by the White House, Codecademy is devoted to teaching people the fundamentals of coding to develop solutions and create business models at no cost. While other internet coding courses are a "learn at your own pace" condition, Codecademy propels students to keep a quick pace by making connections with people in supportive groups and then use a gamified framework. The school offers seminars on various dialects including PHP, Python and Ruby with numerous advantages for students to build upon existing projects before the completion of their course.

**Bloc:** This platform is focused on web development for students who want to learn very quickly. Rather than short courses or addresses, this very organized program keeps running for 25 hours weekly, on a monthly basis. With educational cost beginning at $4,250, bloc.io is on a higher side than the other platforms we have looked at, yet it offers an extraordinary choice for the people who are prepared to focus on a lifelong change.

**Iversity:** This platform is considered to be the "Coursera of Europe". Berlin-based diversity has collaborated with European and global universities to offer academic courses at no cost. Although Iversity is yet to give certification for courses on their platform as opposed to Coursera, which offers certification with its partner universities across the world.

**Skillshare:** This platform uses a different approach as its model is based on a community market to build new skills. With an expansive scope of various subjects to look over, Skillshare offers an online list of video-based courses, just like in-person workshops in urban communities like San Francisco and New York. Numerous classes are accessible to take without enrollment at an expense of around $20-$30 each, yet top classes taught by industry pioneers are only accessible with a Skillshare subscription cost of $9.95 monthly.

**General Assembly:** Focusing on training in design, business, and technology, New York City-based General Assembly has campuses in almost twelve distinct urban areas around the globe. In spite of the fact that most of General Assembly classes are face to face, they offer an alternative of an online-only or mixed blend of teaching methods for students who are registered on the platform. General Assembly even live-streams well-known lectures, with real-time interactions between students and lecturers. Their online courses run in cost, from irregular lectures to multi-part workshops.

# CHAPTER 17

## HOW TECHNOLOGY HAS TRANSFORMED SECTORS

Technology has practically taken over almost all aspects of human activity in both developed and developing countries of the world. The degree to which individuals depend on technology has made it simple for some to do tasks that, up to this point, would have taken a lot of our time to execute effectively. Obviously, technology has been compelling in all areas of the human undertaking; the development acquired through technology can't be overemphasized. It's stunning to look back and see exactly how far our reality has evolved with respect to technology. If someone from 100 years prior, all of a sudden, ended up in this day and age, they would think they had been teleported into another realm. The same can be said about the impact of technology across the board, industries, and sectors of the economy, stakeholders, consumers, manufacturers, service providers, and any other factor you can consider.

HEALTHCARE

Having documents scanned can indeed lower production time, increase communication and collaboration in your team, increase accessibility, free up office space and most importantly save you money! We've covered that already. However, these are just a small number of the bonuses that document-scanning and digitization can offer to businesses. These are just one of the benefits that are accessible to the healthcare industry when digital transformation is involved.

Access and Storage

Archiving and storing all of the patient charts can be a cumbersome process; for instance, from finding the space to storing them, to differentiating between active and inactive patients, dealing with misfiled or misplaced records and, of course, having to physically look for and extract them.

On the other hand, if these charts were to be scanned and uploaded into an EMR (Electronic Medical Records) system, the hospital or medical facility would be able to access them quickly and always store them correctly, and at the same time, not have to find space to house them.

With more patients to see and ever-growing demands, the medical facilities can't risk falling behind on their duties. If a patient sees medical personnel and later needs an answer to a question or demands to see their medical records right away, it could be a challenge at times.

Some staff would have to stop what they're doing, go down to records or archives to find the patient's records before copying them then delivering to the patient. But what if they had an EMR system? The files would be found with just a simple search and emailed to the patient in minutes. This allows the medical personnel more time to do the more important work while still keeping the patients happy, which is a win-win.

Companies like IBM Watson are working out psychological applications that rapidly produce a rundown of potential treatment alternatives for certain health issues and diagnoses. This is accomplished by utilizing AI to break down information in clinical notes and reports with applicable data, for example, clinical ability, extensive research, and information. Whenever improved, "psychological applications" will expand the efficiency of doctors and help improve their efficiency. From improved operational proficiency to guidelines for the treatment of patients, the technological transformation in healthcare has upgraded the whole experience for both medical professionals and patients.

One of the greatest things that technology 4.0 has brought is the availability of information and the way it is stored and processed easily. The Internet, intranet frameworks, search accessibility, and the capacity for healthcare experts to quickly share data have improved the accessibility and examination of health-related issues. "Big data" in healthcare enables the whole field to engage in better examination and analysis processes. This helps bigger and more differing populace that is limited to adequate healthcare than any other time in recent memory. They can likewise draw from existing examinations for thorough meta-evaluation. This technological transformation enables medical experts to remain trendy with patient care patterns, better approaches and innovations. This can be helpful to naturally distinguish risk factors and prescribe the correct treatment by evaluating patient data with that of information from a large number of other patients that might have similar symptoms.

Technological advancements in healthcare have encouraged much smoother correspondence within healthcare service providers. Medical experts will now be able to make use of media like video, online dialog forums, and continuous data gathering abilities to impart and propel the spread of learning in the field. Electronic medical records are now made available to every stack holder and care supplier. This helps to show the progress made in medical issues management, medications and patient recuperation.

Already, medicinal data from visits to the General Practitioner (GP), therapeutic pro, partnered medical experts, and the dental specialists are held in independent areas with various health professionals and clinics. Electronic medical records permit every single patient history, test results, analysis and applicable data to be put away in an online area. The information considers progressively engaged and precise consideration, just as the capacity to see healthy patterns for every person. Technology in medical frameworks helps clinics, hospital facilities, and healthcare practices to function much more easily.

Telemedicine/Telehealth services like video-conferencing are starting to become financially savvy approaches to supplement nearby wellbeing administrations. It is especially helpful to those living in countries, local, and remote networks and requiring essential access to medical services that are not readily accessible to them. For the most part, you have doctors, medical professionals, nurses, maternity specialists, health workers, and practice attendants giving up close and personal clinical administrations to the patient during the teleconsultation while the specialist is making sure that the right procedure is done through conference communication. Teleconsultation is likewise valuable to healthcare workers on the field, where a medical expert can give instruction and training online.

Healthcare applications (and other valuable mobile applications) are vital to improving the correspondence between patients and healthcare experts. These applications empower individuals to handle their health issues effectively; everything from inciting them to getting checkups, to discovering general medical data or getting their test outcomes safely online day in and day out without booking in a meeting with their doctor and waiting for days to see results of medical tests. Healthcare experts can now access more information about identifying certain sicknesses and medications, pictures for clinical issues, and instructions to improve services provided to patients.

Neopenda helps improve care of newborns in resource-constrained hospitals. Neopenda continuously monitors newborns' key vital signs and alerts attending healthcare professionals when a newborn is in distress. The start-up received award funding from Cisco in April 2016 and has since then, received more than half a million dollars in additional funding from other investors to make their first field deployment in Uganda a reality.

The positive effect of technology in healthcare is clear. Healthcare service providers that make advanced progress take advantage of healthcare technology, trends, and development. Adapting to this technical transformation is important for both patients and healthcare service providers. Grasping advanced healthcare services encourages: cutting-edge computerized platforms, improved operational productivity, integrated way to deal with patient medical issues, automated services and clinical procedures, easier cooperation, higher HIMSS, improved possibilities for development and innovation, better patient results and reduced expenses for both patients and medical service providers.

## HOSPITALITY SECTORS

As an industry, the profitability of a hospitality organization depends on a natural and transforming association with the customers it serves. Quick technological advancements have transformed our general surroundings for them to provide personalized services to customers. Key players in this industry are definitely mindful of the role that technology plays in the lives of their clients – and, transitively, in the achievement or disappointment of stakeholders in the industry.

The hospitality industry is all about envisioning and addressing the needs of customers even before they inquire, adjusting to developing industry, and perceiving how technology continues to mold and transform the hospitality industry. With innovations in technology consistently expanding, each part of our day to day lives is being adapted to it, regardless of whether you think technology is changing the hospitality business and improving customer satisfaction or not. Customers are generally expecting quicker, more productive and ideal services from every encounter with service providers, regardless of whether that be a little family eatery or a major chain of hotels, the level of transformation is experienced across the board. It is significant that we adjust to these new conditions, and guarantee that clients are getting a similar standard of administration, regardless of whether you surrender to the specialized insurgency or something else.

Social media, on the one hand, has totally changed the way hospitality service providers relate with their customers, with the business having the option to publicize their administrations in inventive ways and clients ready to impart their experience to the whole world on the web. It is basic that your business utilizes this to further its potential benefit, which if not managed appropriately, can have some lethal repercussions for your business. It's an irregularity these days to have a paying client visit a hotel without having checked

TripAdvisor heretofore. Think about this as a platform to associate with your customers and grandstand your thought. Dealing with your online surveys will allow you to offer pay to miserable clients and guarantee that you are effectively accepting their input. Try not to consider online surveys as a simple negative space to control, as you'll have the option to advance your positive audits that you get to balance them. In any case, remember that it doesn't easily fall into place for individuals to post about an administration that was essentially 'great.' You should go well beyond to guarantee that individuals set aside the effort to compose a decent audit and furthermore brief them during their visit.

Technology is used in many aspects of services provided to customers, one being check in and check out. This is often a lengthy process that most visitors would rather do without. With the aid of smartphone apps, many hotels offer customers the option to check in and check out on their phones, eliminating the need to queue and waste time once they have arrived at the hotel. Some hotels have gone a step further and allowed the smartphone to serve as the key to the rooms once they have checked in, meaning that customers can come straight to their hotel room without interfacing with staff.

Although this ticks the speed requirement, it does eliminate interaction with staff upon arrival to the hotel, meaning that you will not have a chance to promote specific features or interact with your customers. There are positives and negatives to every technological feature you put in place; it's important you recognize them and work out the implications. Many of the negatives can be counteractive; for example, there are many advertising opportunities on the app and potential to introduce a live chat or telephone service. After all, if there is something your customers want, they would not hesitate to call down to reception.

With the use of technology with hotel customers, comes a lot of data that will enable you to home in on your targets and personalize their experiences. Technology is making it simple for customers to specify any specific needs or special requests they might have. If your hotel knows that your customer has a special interest in fitness, you could promote your gym access to them or offer them a special offer. Or, if you knew someone was gluten intolerant, you could have a specific menu sent to their room or a sample of some of your gluten-free foods upon arrival.

With an increase in recruitment being handled online, the process has never been simpler. You can even completely eliminate the need for HR to recruit at all by using a recruitment app or online platform. Many of these apps remove the need for you to filter through candidates and find the perfect match as they do all the hard work for you.

During recruitment processes, with the use of Artificial Intelligence, it assesses candidates to allow employers to see those that are a good fit for the role. All you have to do is swipe right to connect with a candidate; an interview can be set up as early as the same day. By minimizing the time you spend looking for candidates, this eradicates the wasted time spent interviewing people that may not be a good match for your business. It also removes the need for your staff to spend time looking for new hires for you; all you have to do is upload a job role on the app and let the app do the rest.

Technology is changing the hospitality industry in every aspect. You may feel that customers are becoming expectant to be able to navigate their entire stay through an app without having to interact with staff or be able to control the lighting in their room with their phone. Although features like this are becoming more and more common, so are the benefits for the businesses themselves. Technology is enabling you to find out about your customers and support them, using as little time as possible. Although technology is able to enhance your customers' experience massively, it will never replace the human element that comes with customer service.

## HOW TECHNOLOGY IS CHANGING MANUFACTURING

There is no questioning the transformative impacts of technology progressions on organizations. The primary improvements that influence manufacturing are the Internet of Things (IoT), which depicts the manner in which appliances utilize the web to connect and integrate with artificial intelligence (AI) programs, which incorporates huge measures of information for machine learning. Both these headways have improved manufacturing frameworks and efficiencies and are generally welcomed in the segment.

Before the end of 2020, there will be more than 1.3mn robots on production lines all around the world. Nonetheless, the rapid pace of technology advancements is joined by worries that robots will render assembling jobs repetitive and outlived. I trust the opposite - the advancement and execution of these advances will open numerous entryways for organizations, with human and robot coordinated effort being at the front edge, as this pattern is set to proceed, it merits taking a look at the manners in which these advancements are on a very basic level changing the manufacturing business as we probably are very much aware of it.

The goal of every manufacturer is to produce flawless products; this goal sometimes could be almost impossible. With the expanded execution of AI in the production line, it is currently a lot simpler for organizations to compete in the level of productivity, lower error rates, profitability, or efficiency.

Through IoT innovation, machines can consistently communicate with each other and respond to any issues that emerge. When a machine detects an issue, it can rapidly alarm different machines and people, thereby enabling the issue to be attended to immediately. Machines can identify deformities missed by the human eye within a nanosecond. The improvement in precision by technological frameworks reduces the level of errors and risks in the manufacturing sector.

A reasonable case of how the business has been affected by AI and IoT is its effect on predictive maintenance. The part experiences machine breakdown and glitch: an ongoing report by The Wall Street Journal found that impromptu personal time because of breakdowns in gear cost organizations in the US amounts to $50bn every year. Technology Innovation has the ability to make these costly intrusions a relic of days gone by.

Technology gives a framework that continually learns and assesses how a machine is running dissecting information and minor moves in execution. The technology will probably figure out when an issue might happen quicker than any human. In doing such, AI advancements can find any issue at a prior stage, so organizations can predict the issue before it intensifies.

As indicated by a McKinsey report, organizations that have officially executed predictive maintenance inside their manufacturing plants have seen a breakdown of machines decreased by 50% and support costs diminished by somewhere in the range of 10% and 40%.

In the next couple of years, we will see the expansion in the utilization of predictive maintenance, with organizations receiving the benefits of lower maintenance costs and improved efficiencies.

IoT and AI are on a very basic level changing traditional client commitment, a basic region for the industry. The improved administration offering can help in calming grieved or annoyed customers by making clear communication between manufacturing organizations and their clients. Updates on customer services could be made remotely at any point through IoT, and if any product comes with deformity, customers can be quickly informed about the issue. Most times, when customers buy a product with factory error, they would most likely never patronize that brand again to avoid repetition of products with deformity from a particular brand.

## TECHNOLOGY HAS MECHANIZED AGRICULTURE

Technology has mechanized agribusiness: Modern agricultural innovation enables few people to develop immense amounts of food in a brief timeframe with less input, which results in significant returns on investment and a high degree of profitability. Through government subsidies, little and medium-sized farmers have figured out how to get harvesting, sowing, plowing and watering furrowing, through machines. The utilization of technology in agriculture has additionally brought about the assembling of genetic yields. These crops can grow very fast, and they have high resistance to diseases and pests. Farmers have the accessibility to artificial fertilizer, which increases the value of the soil and generates great harvests and empowers them to deliver excellent yields. Farmers in dry regions have been in a situation to develop sound yields; they utilize advanced water sprinklers which bring water from streams to the farms, these procedures can be automated in order to avoid wasting time on a manual procedure. Some countries have advanced technology in their agricultural sector; a case study is Egypt, which is a desert nation and gets little rain, yet farmers have utilized automated sprinklers to flood their farms. In Egypt, they grow a ton of rice, yet this yield needs adequate water to develop well. The water is siphoned from the Nile River to the rice fields regularly.

Technologies based on cognitive IoT are building farms that utilize the practice of smart agriculture by enabling them to collate data from different sources, such as authentic climate information, soil nutrition sensors, advertise data, social media posts, crop pictures, dampness level sensors, and so on that furnish farmers with significant information that will enable them to improve harvest yields.

Harvest quality and profitability, just as asset preservation, are only a portion of the advantages that can be gained from using constant and recorded information in the mix with machine learning algorithms. Cognitive crop framework can search for patterns and yield recommendations for over/under watering, bug control, and soil content alterations.

By expanding harvest yields, this innovation will help battle the issue of global food scarcity with diminishing agricultural lands.

# INNOVATION HAS IMPROVED TRANSPORTATION

Transportation is one of the fundamental zones of the technological movement. Both society and organizations have profited greatly from the advancement in this industry. Transportation gives mobility to humans and goods. Transportation framework is similar to every other technological innovation. It is a progression of parts that are interrelated. These parts all work together to meet a specific objective. Transportation utilizes vehicles, trains, energy, planes, motorbikes, time, people, streets, cars, data, materials, and finances. Every one of these parts referenced work together to move and migrate people and products. Innovation has helped in propelling all the four kinds of transportation and these incorporate: street transport utilized via autos, air transport which is utilized via planes, water transportation which is utilized by boats and speed vessels and space transportation used to go to the moon. The most utilized of all these is Road transportation; this one encourages the development of products and people. Innovations in cars, transports, and trucks have improved the manner in which people move and how they transport their products from one spot to another.

Additionally, developing nations are getting assets from developed nations to improve their street transport, which has brought about the advancement of remote rural and urban areas around the world.

Technologies like automobiles, buses, and trucks have improved the way humans move and how they transport their goods from one place to another. Also, developing countries are getting funds from wealthy countries to improve their road transport, which has resulted in the development of remote rural areas.

Self-governing vehicles, which should fill our roads throughout the following decade, are controlled by AI. A less discussed precedent is the driverless whole deal trucks being worked by the Tesla and Daimler's of the world. The quick open door for self-ruling trucks might be through using sensors to interface numerous trucks by means of IoT to travel together. Self-ruling trucks take heading, impart, and pursue the lead of a truck with a human driver. With this model, a single driver could deliver various truckloads of products at the same time. This arrangement will reduce the expense and speed of delivery for most organizations.

# TECHNOLOGY HAS IMPROVED EDUCATION

Technology has improved the training and learning process. Education is the foundation of each economy. People need well-structured instructive foundations with the goal to read, write, and also figure out how to analyze data. Various schools have begun incorporating instructive advancements in their schools with an incredible point of improving the manner in which students learn. Advancements like brilliant whiteboards, computers, smartphones, iPads, projectors, and internet are being utilized in study halls to support visual learning, as this has demonstrated to be the best technique for learning in numerous subjects like arithmetic, material science, science, geography, financial matters, and considerably more. Some organizations have put in financial incentives, which can be leveraged by educators and students. For instance, on iTunes, you will discover numerous educational applications which can enable students and instructors to trade scholastic data whenever; this has made learning accessible around the globe. Blockchain technology has enabled education database whether you're moving to another school, employment, or nation, a safe online archive could help approve qualifications of previous records. Blockchain could open up enormous open online courses by carefully designing a framework to approve different courses, tests, and assignments that have been finished for a given class. These accreditations could possibly open the universe to authorize advanced education internationally and in developing countries.

## FINANCE AND INSURANCE

An industry that has seen tremendous developments as of late is the financial sector. The new trendy expression Fintech is gradually evolving; with a consistently developing corporate and customer focus, the need to stay ahead in the improvement of customer experience is significant in today's digital world. In a report gathered by PwC, 77% of financial organizations will do their best to increase innovation because almost every business today has an idea of disruptions in Fintech.

The greatest way that Fintech is changing the financial sector is through customer service. Before, good customer service was imperative for any organization associated with finance. Anything that included money-related transactions required staffs to attend to the customer. Today, chatbots have quickly turned into the standard for customers to collaborate with. An AI which advances and gets more intelligent is something which is great on paper, yet in principle, the two come up short on human touch and renders many people surplus to necessity. What is the need to pay staff when the machine works for less? Banking was generally something that was done in the non-virtual world. People would go into town to their bank to withdraw cash; these sorts of premises are quickly becoming excessive. Internet banking is getting increasingly more modern and accepted. We can now easily do financial transactions with the push of a button on our device. We live in a time where we can get to our financial balances on phones, computers or tablets. This is the sort of thing that is disrupting the financial industry and is one of the greater effects on the businesses and customers.

Banks are not just a part of our lives, but they also play a significant role in our daily lives. For many of us, the day cannot end without at least a single financial transaction taking place.

Thus, banks always try to adopt the latest technologies in order to enhance customer experience.

Digitization cannot be seen as an option for banking industry but rather an inevitable phenomenon because every industry is being digitized and the banking sector will be no exception.

In fact, the use of mobile banking is increasing at a much faster pace than online banking.

Here are some of the advantages of digital transformation in banking:

Improved customer experience.

There will be reduction in costs for the customers and the banks themselves by the use of ATMs and cashless transactions.

With more digital data available at the banks, they can take data-driven dynamic decisions by using digital analytics.

This benefits both customers and banks.

Technology is non-discriminatory and so everyone will be treated the same at banks.

The number of customers at the banks will be increased because of the increased convenience of banking.

Digitalization definitely reduces human error and by extensity, the instances of handling large amounts of cash will be drastically reduced.

Opening and maintaining bank accounts will become even easier, and repetitive tasks will be eliminated by the use of automation.

Rural and urban gaps will also be eliminated.

With the increase of cashless transactions, the use of fake currency will be hampered.

AI is advancing past the limit of human expectation, a machine can follow through the historical backdrop of the person in question, and after that figure out and anticipate the probability of misrepresentation dependent on past crimes with more speed than a human could.

Generally, these are only a couple of the various ways that innovation is modifying the manner in which the financial sector works. The essential interruption originates from advancement. Advancement in any field is typically a type of disturbance. Disruption in the financial sector is happening on the grounds that the current practices are rendered outdated, and should be redesigned or expelled. Technology will advance to the point that it becomes more proficient than the general population that made it. There's a convincing case for a world dependent on machines. They're more brilliant, quicker, less inclined to oversights, and substantially more financially feasible.

## RETAIL

In the retail business, technology is changing the manner in which numerous parts of the business are directed. Until all around as of late, customers have needed to remain in long queues to buy their products. On account of technology, however, customers can search through any online store from their device without visiting a physical store. This has, altogether, improved the satisfaction of customers. Before, about 10% of customers would leave without making a purchase due to the fact that they cannot stay in a long queue. The utilization of innovation in the retail sector has enabled payments by installments feasible. This innovation has disposed of the requirement for passwords or PINs and has streamlined the checkout procedure for clients. Quicker exchanges have prompted higher deal volumes in stores that offer remote POS and contactless installments.

Machines such as self-help bots are a fantastic utilization of IT in the retail business. Innovation has likewise improved the manner in which people shop online by giving a customized shopping background. Technology is additionally being used by organizations to offer their customers virtual perspectives on products through augmented reality. Customers are able to feel what they want to buy before making a purchase as a result of augmented reality in the retail sector. With this innovation, organizations have improved the experience of their customers and expanded their deals.

From the retailers' perspective, the act of overseeing stock has dependably been all around expensive for organizations. With the innovation that can track stock through its purchase cycle and offer continuous data about the product to the board, stock administration has shown signs of improvement and is costing organizations much less. With technology following things and offering ongoing updates about the things, directors have a vastly improved feeling of what is being obtained and what things they have to request to keep their product supplied. At the point when technology is utilized to monitor stock, store product is increasingly sorted out, and the potential for worker burglary is definitely diminished.

Much the same as with stock administration, value evaluating has been a tedious and expensive procedure for organizations. Value reviewing is fundamental for organizations to guarantee that they're not cheating or undercharging their customers. Innovation has streamlined this procedure via automated value checks when products are filtered. This makes room for precise evaluating, spares store workers a great deal of time and builds stronger trust between the store and the customers. The effect of this data technology in the retail sector has been extraordinary and advantageous.

## AVIATION AND DEFENSE

The world is, in fact, changing as a result of technology. As cutting-edge innovations become less expensive and accepted, new regions are being entered where the physical and digital universes cover and mix together.

Organizations accept these developing advancements not exclusively to digitize their activities but to take an interest in the adaptable and always captivating business ecosystem. The aviation and defense (A&D) industry has been among the early adopters of advanced technology. A&D players have been utilizing automation and robotics in their sequential assembling line for the last few decades.

Organizations have a tremendous spotlight on utilizing propelled manufacturing methods to reduce costs and improve profitability. The gigantic measure of information accessed through sensors and analytic investigative strategies has opened up a view into the insights of activities around product performance, operations, and customers that were never investigated. A&D organizations are under consistent strain to expand and improve operational effectiveness; they have to adjust to business interruptions. The utilization of 3D printing technology is to tweak airship and defense machinery for various customers. They send huge 3D printers to manufacture durable parts from heavy metals like titanium and tungsten with very little wastage. These automated printers offer extraordinary accuracy and low maintenance. Machine learning empowers suppliers to persistently screen and analyze huge volumes of information identified with machines to distinguish oddities and foresee breakdowns. Production of a digital twin, a virtual replica of the machine in the cloud, enables the advancement of tasks for different results. Cognitive experiences empower to improve efficiencies and configuration of product designs. Continuously, using AI and machine learning makes autonomous frameworks, which would act naturally adequate in setting up flight operations and front-line choices all alone.

Virtual and augmented reality technologies are disrupting the fix, maintenance and operations by giving designers more data about basic parts and visibility to support maintenance. Augmented reality innovations can make simulations for pilot and defense force to increase their experience through utilizing virtual A&D frameworks before even physically handling a plane.

## UTILITY AND ENERGY

Generally, utilities have had little involvement with rivalry or techniques for winning new customers. Union and globalization have changed that, as the business changes into an increasingly powerful market-driven environment with less, bigger organizations in this sector.

Innovation has opened up new chances and new dangers, constraining utilities to grow their concentration from essentially giving energy, to reexamining their methodologies and adjusting them all the more to meet up with customer needs. Utility suppliers are, in this manner, utilizing data innovation and automation in their business operations to improve operational productivity and keep up with a growing customer base.

Globalization has prompted progressively open markets, empowering organizations to enter the new market more effectively, yet presenting them to more prominent challenges. Privatization in certain countries has additionally brought about the age of new capital, permitting further interests in the framework and inventive innovation. The market restructuring has prompted more customer decision, aggressive evaluating, and new customer structures, driving utilities to overcome various pieces of the value chain as top tier suppliers.

Customer Information Systems (CIS) for utilities have been around in some structure for over 20 years. In the late 1990s, when utilities initially started to use new software developments to help deal with their customer relationship, the market involved merchants selling mission-basic undertaking frameworks that were concentrated principally on operational customer relationship management (CRM) capacities, such as account upkeep, request preparing, charging, credit accumulation and records receivables, with some community-oriented customer collaboration. In this conventional methodology, organizations delivered costly plans for various procedures. This methodology was set apart by various interfaces, and superfluous information replication, bringing about a surprising expense to evolving markets. As the inconsistency among ventures became increasingly evident, utilities became moderate in their appraisals.

The present organizations are currently looking for arrangements that will enable them to streamline process effectiveness, develop techniques for adjusting change, speed time reduce administration costs, improve overall revenues and drive investor confidence.

Customers are the foundations of business in the utilities and energy industry. The progress achieved by deregulation has become most obvious in this sector. Deregulation has expanded the requirement for escalated information trade and better coordination of organization and bookkeeping practices in this industry.

## CRYPTOCURRENCY

The very essence of technology is to remove pain and chaos as much as possible, the painful and rigorous methods of getting work done, as well as giving direction and order to the way of life. One of the areas of life needing improvement is the conduct of financial services, the economy. Not only for the present, but also for the definition of what the future will look like. One of the plagues of the current world order is the financial institutions. Technology has once again brought an alternative into view which is what we regard now as cryptocurrency. It must be put into consideration that before cryptocurrencies, there already existed digital currencies such as Wei Dai's 1998 proposition of 'b-money' and Nick Szabo's 'bit gold'. If we consider the coinage of the name 'cryptocurrency' we have 'crypto' which has to do with security, and 'currency' which is of course, a denomination of money. We injunct that crypto currency means a secured currency. A digital asset that can be transmitted and utilized in every way possible.

From ages long, the economy of the world had been in the hands of the government and a few individuals who cared little or not at all for the wellbeing of others who were either less privileged or unfortunate as the case may be. Cryptocurrency, which is basically digital currency, was first introduced a decade ago by Satoshi Nakamoto. It is to be noted that before Bitcoin, there have been several attempts at designing and creating a digitally operable currency; however, Satoshi Nakamoto created the first successful cryptocurrency. The concept is a simple one (if judging by the Outlook) which seeks to perform many basic functions.

First of these is to hand the tide of the economy over to the majority, the poor, and the masses. Second on the list is that this new system is also aimed at removing financial institutions from every transaction worldwide thereby removing the exploitative presence of the third party — middlemen. Also, cryptocurrencies are known for kicking against inflation and dropping in the value of

Fiat money. And lastly, cryptocurrency seems to bring about a complete and more secured end to end trade.

After the launch of Bitcoin, many other cryptocurrencies have been launched following the same pattern, albeit with slight modifications and measurement of unit. Top of these list are Ethereum and TBC among others.

Question we should ask is: after the first decade of its existence, has it been successful? Although it is true that many of these cryptocurrencies have failed over time mostly due to mismanagement and other related issues. Despite the fact that there have been some earlier attempts, most have failed, yet, cryptocurrency has proven to be a large technological marvel. Taking Bitcoin as an example, when it was first used, over 10000 BTC was used to purchase just one pizza. And what is the value of 1btc today? It is well worth $5000; that is a huge difference. The same could be said about Ethereum and other successful cryptocurrencies.

Cryptocurrency is powered by blockchain technology which is one of the wonders of the present age. It Can Never Fail.

The difference between digital currency and paper money is that while the value of the latter falls, the former only adds more value. A dollar's value can buy less of what it could be used for ten years before now while I BTC can buy much more than what it could have been used for back then.

*The key to success in cryptocurrency is getting in at the right time*

Since the basic reason behind the invention of cryptocurrency is to favor the common man, this is done in two major ways:

First, the middle man is removed from the transactions being carried out in a cryptocurrency market. Secondly, it opens up a levelled ground for both the rich and the poor. Perhaps, the most known crypto in the world is Bitcoin. But the truth is that there are several other platforms out there with similar or even better working systems.

## The History of Cryptocurrency

Before Bitcoin, there were a few attempts at digital currencies with similar ambitions as Bitcoin, but they were unable to reach the same heights of popularity. Both "B-money" and "Bit Gold" were prior cryptocurrency concepts that incorporated the solution of mathematical problems into the hashing of a blockchain. Bit Gold's proposal, written by Nick Szabo, also involved decentralization.

The first iteration of what has since become cryptocurrency, however, is Bitcoin. And that story begins in 2009, when the entity known as Satoshi Nakamoto created and released Bitcoin into the world. Nakamoto's true identity is unknown; some believe it is one person, others believe it is a group. That same year, Bitcoin software was made public, allowing people to mine bitcoins and create the first Bitcoin blockchain.

The notion of cryptocurrency has intrigued some and turned others off, and the concept has likely baffled even more people. Some think it's the wave of the future while others dismiss it as an online fad. There's a group that believes cryptocurrency and the technology behind it can change the world for the better; there are others who see it as a dangerous trend that wastes energy. In using cryptocurrency for an exchange instead of fiat currency, crypto owners don't have to rely on banks to facilitate transactions, and can successfully avoid the fees that come with using financial institutions.

Generally, cryptocurrency transactions are processed and completed via a blockchain network. Blockchains are designed to be decentralized, and so every computer connected to the network must successfully confirm the transaction before it's able to be processed. Ideally this creates a safer transaction for everyone involved. It can also lead to you waiting awhile; one big complaint about Bitcoin is how long it can take for a transaction to go through.

Cryptocurrency transactions are put into a "block," and the computers in the network get to work solving a complex mathematical problem. Once a computer solves it, the solution is shown to the others on the network, and if the whole network agrees that this solution is correct, that block is added to the chain and the transaction is completed. Multiple transactions in one block makes it harder to edit a single transaction; the network is constantly re-confirming the blockchain on its way to the latest block and will notice should a suspicious edit be made to one transaction in a block. A block is a transaction unit, and many people use the system at the same time.

The mathematics problem here is basically the conversion of unit by unit price of the cryptocurrency.

For example, if 34000 Satoshi (Satoshi is the unit of Bitcoin) is to be transferred between partners, the computer will do the conversion, and the parties will check if the answers are right. When these transactions are made and accepted, it will be added to the chain of transactions that had been done earlier — for the record.

## Cryptocurrency vs. Banks

There are banks interested in what blockchain can do for them, but cryptocurrencies like Bitcoin were developed expressly to avoid the use of banks altogether. Fans and developers of crypto like the idea of a decentralized network that does not require the need of any other parties to process a transaction - and as a third-party with a centralized network, a bank is not where cryptocurrency owners generally want to go with their stash.

Mining cryptocurrency? creating coins, Bitcoin? This could've seen as a novelty in the early days. However, it advanced very rapidly, and it established itself as something that could be used as real alternative currency for the first time, for example in 2010, when someone successfully used 10,000 BTC to buy two pizzas. As of this writing, 10,000 BTC are currently worth more than $109 million. *

See the difference?

This points out that above all things, cryptocurrency seeks to steer against inflation. The more Fiat/paper money becomes useless, the more cryptocurrency becomes more valuable.

There are a few different methods of acquiring a cryptocurrency if you're still interested in getting some. A few of the different places you will be able to find cryptos include:

- Cryptocurrency software
- Cryptocurrency exchanges
- Cryptocurrency P2P (peer-to-peer exchanges)
- Cryptocurrency ATMs

Which one you use will depend on a number of factors. Are you able to buy your preferred cryptocurrency with fiat currency, or will you need to exchange other cryptos for it? How much time and money do you have, and how much energy are you willing to use?

By cryptocurrency software, we basically refer to mining sites. Mining means generation of these currencies through usage of energy. We can generate them with the internet. Let's put to mind that usage of internet consumes energy from the providing servers.

Cryptocurrency exchanges has to do with converting one cryptocurrency to another. Ethereum to Ripple. Bitcoin to Ethereum. Ethereum to Litecoin and so on. Peer to peer exchanges have to do with the buying and selling of these coins between two agreed individuals.

The most popular and common way to buy cryptocurrency is via a cryptocurrency exchange. An exchange is a platform that allows you to trade for or purchase a cryptocurrency. Some allow you to use fiat currency like USD to buy, but for others you may need to already own some cryptocurrencies like BTC that you can exchange for another.

Some exchanges exist as platforms simply to trade - ironically, these exchanges are centralized. In these cases, the exchange offers the convenience of simply buying or selling cryptocurrency and simply takes a fee. Other exchanges, peer-to-peer ones, offer the ability to put you in contact directly with the trade you are buying from. Do research not just to see what exchanges offer what, but what their reputations are; a p2p exchange with a seedy reputation may be a one-way ticket to getting scammed.

Above all, the creation and sustenance of cryptocurrency is another landmark achievement and contribution of technology to the modern world.

**Centralized cryptocurrency vs decentralized cryptocurrency**

In the cryptocurrency world, there are three types of systems in operation. These are: the centralized cryptocurrency, the decentralized cryptocurrency and the new hybrid cryptocurrency. Since the emergence of cryptocurrency, the discourse had been ongoing and with more than a thousand coins in existence already, there are ongoing debates revolving around which system works better.

Centralized companies— even beyond cryptocurrency — are run by establishments and companies that manage the block chain and benefit from the transactions made on the system. There is a singular control of the currency. This singular entity determines the price of the currency and the pace of the transactions. Furthermore, these currencies are only usable within the limit of the centralized authority. The centralized system seems antithetical towards the very essence of cryptocurrency because such system has limited freedom and the third party is once again chiefly reintroduced in the form of the managing companies. A centralized cryptocurrency however has more stability, and less glitches can occur during transactions. But this stability can also be problematic. If the central authority makes any mistake in the process of management, the entire block chain suffers from such a wrong step because the whole system feeds on the central authority. There are also no adequate checks and balances to ensure a fair block chain. On a fairer note, the centralized cryptocurrency is much more secured. Since 2011, there have been over 50 hacks of cryptocurrency exchange systems. This is what the centralized cryptocurrency seems to remove by replacing the public key encryption with the private key encryption.

A decentralized cryptocurrency is one where multiple entities or bodies control the operation of the block chain. There is a widespread of controlling bodies that determine the price and policies of operations. The buyers and sellers of such cryptocurrency are also allowed maximum control of trade up to some stages. The system allows them an even more private identity. The transactions are complete and done without the involvement of the major companies. In short, there is a collective ownership of the block chain. The disadvantage of a decentralized cryptocurrency however, is that there is always a lack of direction and organization. This makes it quite frustrating and tedious to operate. Also, unlike the centralized cryptocurrency, the decentralized cryptocurrency often comes at odds with legality. Bitcoin, Litecoin and Ethereum are examples of a decentralized cryptocurrency.

A hybridized cryptocurrency is one that shares the attributes of both the centralized system and decentralized system. It removes the disadvantages of both and blends in their advantages. The system allows partial control of major shareholders thereby making it conducive for transactions and also stabilized.

According to Wikipedia, the total cryptocurrencies market capitalization is valued at $100 billion which turns out to be bigger than the GDP of 127 countries. And presently, there are over 1600 cryptocurrencies.

## Bitcoin

Bitcoin is the cryptocurrency with the highest market capitalization. As stated earlier, Bitcoin heralded the new age of digital currency and has grown over time. The first Bitcoin transfer was between the founder Satoshi Nakamoto and Hal Finney who downloaded the Bitcoin software on the day it was launched and received 10btc. And coming from the purchase of two pizzas with 10000btc, the world had witnessed the peak value of BTC in December 2017 where one Bitcoin is valued at $19000. The system is decentralized and allows each member a privilege of participation in control of the block chain. However, there is a maximum amount of BTC that can be in circulation owing to the process of digital mining that is used to create it. The limit of Bitcoin is 21 million BTC. Bitcoin is measured in Satoshi as a unit. 100000000 Satoshi makes up one BTC.

## Altcoins

The success of Bitcoin had given a rise to the creation of other coins which are collectively called altcoins. Altcoins are alternative coins that serve the same purpose of peer to peer transaction, albeit with slight modifications. Notable among these are:

- Ethereum
- Ripple
- Tether
- Litcoin
- Robocoin
- Dogecoin among others.

## The Billion Coin (TBC)

The Billion coin is another special and different cryptocurrency which must not be overlooked. This is because this cryptocurrency is created with a definite target, which is eradicating poverty from the world. TBC was created by Kris Kringle in 2016 and since then had gained maximum attention. The difference between TBC and BTC (including altcoins) is that while BTC is market based, TBC is community based. Which means that TBC is not on the coin market. Also, while the prices of BTC and other coins may rise and fall depending on the demand and supply factor in the market, TBC's value only rises. It never drops in value. Beyond being a means of monetary transaction, TBC is a technological revolution movement. TBC's unit is denoted in Kringles.

## Facebook Libra

In June of 2019, Facebook announced its own cryptocurrency called Facebook Libra. This cryptocurrency, which is proposed to be launched in the first half of 2020 will allow transactions to take place with near zero cost. Like DasCoin, Facebook Libra would be a hybridized cryptocurrency. This is because the ownership and control would be shared between Facebook and other investors who have invested into the project's operation. These include: Uber, Visa, Andreessen and Horrowitz. Also, Facebook is launching a branch of its organization that would manage the cryptocurrency transactions. This company is called Calibra. The goal is to create a globally acceptable cryptocurrency using the already broad base of Facebook as a propeller. Also, the value of Libra is meant to be stabilized and less volatile, less dependent on demand and supply. Another modification to the existing cryptocurrencies is that Facebook Libra would be doctored to be a definite mean of exchange, making it more flexible and efficient, thereby posing it at the center of the cryptocurrency market. In fact, Facebook Libra is the cryptocurrency version of PayPal. Facebook is made to ensure that the coins can be used for transportation at any time with little cost as well as anonymous presence. Also, the currency is designed to be easily understood by the average consumer. In summary, it is a new and mainstream digital currency. The purpose of this new currency, however might not be for the uplift of the masses like Bitcoin, altcoins and TBC. Facebook Libra might be another oligarchy when we consider the requirements and the assets of the founding members.

It is to be noted quite importantly, that the buying power of Libra coin although designed to be mainstream, might be limited to the reach of the Facebook community, WhatsApp and other properties of Facebook. Facebook will also be the principal determinant of 'how' transactions are to be conducted.

## CRYPTOCURRENCY AND DIGITAL MARKETING

In the light of the flourishing presence of cryptocurrency in the 21st century, it is only wise for everyone with a foresight to embrace what cryptocurrency has to offer. For a digital based market, and a marketer, what does cryptocurrency amount to and how can it be employed as a strategy of trade?

For starters, every digital marketer should come to terms with the reality that the digital space is quite highly competitive. And one must aim at leapfrogging the existing domains and hierarchies by becoming more efficient through an effective marketing strategy. Your brand presence must be established in order to achieve targeted goals, especially in this new world order.

One very important tool for strategic mastery for any digital marketer is block chain technology. It should be noted that block chain technology is the engine room that powers every cryptocurrency. A block chain could also be explained as a catalog of data in a decentralized system that allows multiple transactions from one end to another at the same time. This technology was brought into existence and improved upon by cryptocurrencies. This technology is designed to render useless or even totally remove some industries that serve basically as middlemen. However, for the digital marketing industry, block chain technology and cryptocurrency make things easier.

Digital marketing essentially means the use of digital technological mediums for the advertising of products. Digital marketing is hinged on the growing and effective presence of the internet since the late 1990s.

According to Wikipedia, "Digital marketing methods such as search engine optimization (SEO), search engine marketing (SEM), content marketing, influencer marketing, content automation, campaign marketing, data-driven marketing, e-commerce marketing, social media marketing, social media optimization, e-mail direct marketing, display advertising, e–books and optical disks and games are becoming more common in our advancing technology. In fact, digital marketing now extends to non-Internet channels that provide digital media, such as mobile phones (SMS and MMS), callback and on-hold mobile ring tones. In essence, this extension to non-Internet channels helps to differentiate digital marketing from online marketing, another catch-all term for the marketing methods mentioned above, which strictly occur online."

To be added to the above list of methods of advertising and selling points is block chain advertising. Block chain advertising removes the blockages posted by Google and Amazon in the process of getting an advert to targeted viewers. It is a direct form of presentation and communication. Although the language of some cryptocurrency is complex, a good digital marketer would build a channel of communication with the audience to not create a void of dissemination.

Block chain marketing is also cheaper in the long run. The smart digital marketer has two main objectives, which are: to make profit and to minimize cost. Block chain advertising runs on a lower cost and could be the thin line between the ordinary digital marketer and the smart digital marketer.

Verification of work done is easier to cross check in a block chain. The advertiser gets to verify the traffic from their end and also send payments at little or no cost depending on the cryptocurrency used. There is a verifiable and auditable record of all clicks on the advert. Block chain would benefit the digital marketer to measure the effectiveness of an advert campaign and also optimize its frequency. Resulting from using cryptocurrency as a digital strategy is that the digital marketer would be more trusted and attract more customers due to the transparency of transactions and thorough process.

Putting this into consideration, employing cryptocurrency and blockchain technology as a digital strategy for a marketer seems like a wise step. And lastly, considering the wild waves that are attached to the move of cryptocurrency, we could be said to be in a financial or monetary revolution and in that case, investing earlier and identifying with block chain technology would give such digital marketer an edge over others in the same industry.

# CHAPTER 18

## SOCIETY IMPACT

Recent years have seen accomplishments in AI using neural frameworks, which copy the techniques of genuine neurons. Rapid technology advancement has made disruptions happen very fast, in a couple of years to come the world will have experienced more innovations in every sector. For seniors, the persistent changes can be difficult to stay aware of. Fortunately, there is a great deal of new innovation for seniors that is explicitly intended to be useful and simple to utilize. Instead of being overpowered, grasp innovation to find how it can upgrade and move toward becoming a significant piece of your everyday life.

Technology has had an amazing impact by giving people all over the world useful information to make their everyday life simpler and more productive. The impact of the new innovation on media is obvious since a media organization is not really a news stage any longer. A media organization is currently known as an organization that helps pass data over the globe. The measure of worldwide internet users is presently close to 3.2 billion people. That is practically 50% of the total population. Around the globe, two million phones are sold each day. The measure of data being shared via the internet is wonderfully enormous; the internet has had an effect in everybody's life. Prior to the coming of versatile innovation, you needed to look through a dictionary to know the meaning of a word. Today, Marvel is utilized less.

Throughout the years, technology innovation has altered our reality. Technology has made astonishing tools and assets, making valuable data readily available. Current technology innovation has made it workable for the disclosure of numerous multi-gadgets like smartphone, smart-watches, robots and so many mobile and digital technologies. Technology innovation has likewise made our lives simpler, quicker, better and more fun; however, this has additionally come with some of its difficulties. The fate of technology innovation is significantly more intriguing than what is going on at this moment. In a couple of years, driverless vehicles might be the standard for everybody, and robots will have a typical spot in careers, homes, malls and the general environment. Today we can connect with people from across the globe through our social media platforms such as Facebook, Twitter, or Skype regularly.

## Technology Has Changed How We Communicate

It is astonishing to think back on how communication has gotten simpler throughout the years. Previously, you needed to compose a letter to speak with somebody. The letter could take days before it was conveyed. You likewise needed to follow the letter to check whether it had been sent to and gotten by the suitable individual. Mistakes were normal also. There is no such marvel today thanks to digital transformation. To connect with people today, there are several choices readily available. You can send them a message through social media platforms, email, messaging apps or put a call through. The manner in which we utilize our phones has changed radically too. There are several mobile applications intended for communication.

Everyone knows about Facebook, Twitter, or Skype; they are good sources for regular communication and are not leaving at any point in the near future. The message gets conveyed on these platforms at a similar rate and speed. It would be sent regardless of whether the message beneficiary has access to their device or not. Also, voice messages through these platforms can be delivered in nanoseconds.

## Technology Has Changed How We Pay Bills and Transfer Money

No more do you need to enter a bank to withdraw cash or transfer it to somebody. Several banks have effectively made financial services available on the internet. Organizations like PayPal have made financial platforms where people can send and get cash from any area utilizing the web. Paying bills has additionally been disentangled gratitude to innovation. Payments by installments has been digitized. With wireless banking applications, a whole lot of financial transactions can be done on mobile devices. It's even simple to pay for things utilizing just a button on the smartphone. You could leave home without your wallet and be okay. Financial trends have given rise to a cashless economy in some countries.

## Shopping

Digital transformation has changed the way people purchase what they want.

Through digitization, e-commerce has revolutionized the majority of customer attitudes towards shopping. One can just click a few buttons from the comfort of their living room and forgo the necessity of stepping into a store to shop for the goods they need and desire.

Apart from enabling you to accept the delivery of your preferred purchase to your doorstep, benefits like making online payments or exchanging goods also help in promoting the online shopping craze.

With this technological innovation garnering traffic and use from different sections of shoppers, here are ways that digitization has impacted the future of online shopping.

A plethora of shopping options.

Unlimited and unrestricted choices have become the catch word for attracting online shoppers to spend more money through digital interfaces.

In an attempt to encourage even more patronage, online shopping interfaces will keep stored information as part of the shopper's digital fingerprint.

This will help marketers offer products that are desired by their consumers.

Shoppers will also experience transparency and a sense of customer-friendly service which in turn, will do away with the need to repeatedly verify their identity and subsequently, their card number for every transaction.

Browsing history of online shoppers – a treasure for marketers

The browsing and shopping histories of online shoppers is just like a wealth of knowledge for online marketers, who have, as their main objective, obtaining such data.

This, when teamed with a detailed listing of a customer's basic data and said customer's recent transactions and interactions with customer care professionals, will greatly highlight the purchase behavior of customers.

There are various ways to comprehensively analyze customer spending habits. When this information is captured by specialized technology from brands like

Lintech or online retailers, marketers will enjoy higher payoffs through repeat purchases, while shoppers get exactly what they want without having to start their search all over again.

With the introduction of e-commerce, people with busy schedules can still find great sales deals on e-commerce websites which has become an enormous online market place today. There is no compelling reason to make a few treks to retail stores with sites like Craigslist, Amazon, Facebook, and Instagram Marketplace where a lot of commercial activities can take place online with a connected device.

### Education

Digitization has without a doubt, changed our education system but we can't say that it has diminished the value of our traditional classroom learning nor do we want something so priceless to turn into dust.

The best part about the digitization of education in the 21st century is that it is combined with the aspects of both classroom learning and online learning methods.

Both aspects act as a support system to each other which presents a unique strong point to our modern students.

Digitization in education has also proven to be the right method for saving resources as far as impact on society goes.

Online examination platforms have restricted the excessive use of paper, directly reducing the need to cut down trees. This way, the digitization of the education industry in the 21st century proves to be a boon to our society.

**Technology Has Changed How We Date**

The internet has improved our lives tremendously. A whole lot of relationships have been built one way or another online; one route is through the simplicity of people connecting with different people across the world. With the expansion of internet, dating sites have been able to match people based on their preferences in finding a partner. Digitization has made it possible to run some background check on a prospective date before meeting them in person. Some analytics can be run on a potential date to determine their likes and character.

People can now access the data needed to think about a specific individual. Particularly for seniors, internet dating is useful during a phase of life where it's ordinarily hard to meet new individuals. Dating platforms have helped individuals interface with their match in a manner where neither one of the parties needs to waste the other individual's time. It is agreeable, clear, safe, and it works. Dating platforms have made finding love easier and faster based on predictive analytics in matching potential partners.

## How the Internet of Things Works with Home Devices

The majority of the innovation referenced above has been made conceivable as a result of the internet. Platforms, new sites, and mobile applications will continue to be created to alter the manner in which we live our lives. The most recent technology innovation is not yet available on screens: the Internet of Things (IoT). This enables ordinary items to be synchronized to the internet through Wi-Fi. This makes their functions open remotely and automated through information accessible on the World Wide Web. IoT today incorporates driverless cars, wellness trackers like FitBits, indoor regulators and doorbells. Since these gadgets are smart, little human effort is required. The sensors on these indoor regulators are explicitly intended to create a smart home with ideal temperature. The indoor regulator can give the best temperature. However, when the gadgets connected to the Internet of Things appear to be overpowering, control is required. There might be a simple arrangement: voice collaborators. These incorporate Amazon Alexa, Google Assistant, and Apple Siri. With a voice aide, control of gadgets in the home is possible online and offline.

## Technology for Food Delivery

A great deal of the innovation we have effectively portrayed helps make our lives simpler by sparing time and resources. Another case of this sort of innovation is food delivery. The capacity to have food delivered to homes is particularly useful for seniors or those with very busy schedules. One part of our lives that we frequently underestimate is the ability to go to the supermarket. Right now, there are numerous mobile applications that enable deliveries from preferred cafés, restaurants, pizza places, grocery stores or bakeries. These applications pay individuals to get takeout anyplace. A couple of instances of these applications incorporate Postmates, Uber Eats, and DoorDash. A lot of these applications offer basic deliveries, for example, Instacart and Shipt. While Walmart, Target, and Kroger even offer their own food deliveries. It is no longer difficult getting breakfast, lunch or dinner as they can be delivered to any doorstep with the click of a button on a connected mobile device. For those who don't cook, there are additionally organizations that deliver special diet plans for any meal. A few instances of these organizations incorporate Magic Kitchen, Silver Cuisine, and The Good Kitchen. Despite everything, they taste incredible in the wake of being warmed yet likewise contain healthy benefit for various individuals, including veggie lovers, diabetics, and seniors. These alternatives likewise reduce the tedious shopping for food and cooking.

## Technology in Transportation

We addressed driverless cars and the innovation behind them (IoT), yet this innovation presently is not yet commercialized. When it is, it will prompt fewer mishaps and accidents caused by human error. It will likewise make transportation progressively open to everybody. Driverless cars will utilize sensors to know when to break, accelerate, turn, and park. Until the advancement of innovation for driverless cars is safe and good to go, a larger number of people will still use planes, trains and normal cars to travel. Innovation has made a portion of these alternatives simpler to utilize. Today in the sharing economy, a ride can be shared with someone going your route. Companies like Uber and Taxify have made this possible. The innovation in their applications enables people to pick a ride from anyplace, much the same as a taxi. The best part about these applications is that the driver is identifiable from a GPS map. Rides are open even in zones where there are no taxicabs. Organizations give people the opportunity to become partners by becoming drivers. Regular individuals with modern cars can apply to become drivers. When they pass the verification test, they become qualified to begin giving rides through the application. A few people are reluctant to take rides from outsiders. To make it more secure, the applications have rating frameworks set up to guarantee the drivers are carrying out their responsibilities well. The organizations likewise constantly update their safety efforts to ensure riders are protected.


## Healthcare

Through innovation, we are living longer than at any other time. There have been lots of healthcare innovations in social insurance, medicines, and hospitals; these upgrade our regular day to day existences with security and assurance. Security gadgets at home can protect people from hazards. At the point when these cautions go off, they can start recording with a camera around the home, and this will alarm security experts. The interface with a healthcare application can be integrated on a smartphone. Healthcare specialists need to review data physically before they analyze or treat a patient. Artificial intelligence can help in therapeutic services by diagnosing patients faster and even more accurately, making innovative new prescriptions and medications, thereby lessening medical errors and lowering the costs of social insurance for providers and patients.

**Virtual Assistants**

A menial helper, otherwise called AI colleague or advanced aide, is an application that integrates programming language and voice directions and finishes assignments of clients. Siri, Google Now, and Cortana are generally clever progressed digital partners on various platforms (iOS, Android, and Windows). They help us in finding supportive data or information when we request that they utilize conversations or voice directions. These partners react by discovering data and getting results.

**ADVANTAGES OF TECHNOLOGY**

It helped generations from industry 1.0 to 4.0. Innovation has essentially improved proficiency and profited organizations in all sectors across the world. From automakers to cooks, everybody is making the most of its advantages. Organizations these days can undoubtedly achieve their objective and satisfy customers at a much-decreased expense with the use of the Internet. Data is now readily available. Searching for a fix to a migraine? Need to do some shopping? Composing a scholarly paper? Hunting down an occupation? You would now be able to do these without going out. The Internet specifically empowers us to extend our arms of friendship globally. We presently can communicate through various medium like Facebook, LinkedIn, Google+ or Twitter.

## DISADVANTAGES OF TECHNOLOGY

It has brought forth damaging weapons. Innovation has given us the nuclear bombs, atomic weapons, poison gas, cyber threats, addictions, anxiety, social disorder, eating disorder, obesity, pollution, social isolation, cyberbullying, privacy issues, eye problems, bad posture, and several others. One late development that could simply crash the entire of mankind easily is 3D genomic printing. This innovation enables anybody to print any sort of thing available to him or her. It has many negative impacts on people. Innovation, especially as PCs, advanced mobile phones, iPods, long range informal communication destinations and so forth, present much damage to a distinctive individual in both the mental and physiological perspectives.

**Social Isolation** We have figured out how to stroll with our iPods and smartphones without physically communicating with people within our environment. **Absence of Social Skills** Non-verbal communication and expressive gestures in people is gradually reducing as they tend to keep an emotionless face.

**Absence of Privacy** Gone are the days that we needed to approach an individual in an event to get their number or address. With a couple of searches, anybody can get one's location and contact data without much stress.

With digitalization disruption, one of the key downsides is the loop holes provided by this revolution that allows certain security compromises to occur in the business world. Cyber-security and privacy have been a major threat to the many benefits brought to light by digitalization, and these are really disturbing considering that these downsides are part of the foundations for the future of work, business and even safety of end-users. It is important that we dive deep into what cyber-security and privacy rights are all about. The next chapter will be shedding more light on the issues of cyber-security and its impacts as a factor responsible for digitalization.

# CHAPTER 19

## CYBERSECURITY AND PRIVACY

As I meet with government, business, and technology leaders, the single most important challenge that they must all confront is cybersecurity. They not only see it as a challenge, they see it as a very enormous obstacle that is hindering the progress of digital transformation. According to a recent study, the cybercrimes damages will reach more than 5 trillion US Dollars by the year 2021. And here are some astounding findings from the Cisco Cybersecurity Annual Report: 42 percent of executive leadership consider cybersecurity high priority, 95 percent of all attacks on enterprise networks are the results of successful spear phishing and 27 percent of advanced email attacks are being launched from a compromised email.

Cyber-security incorporates a variety of difficulties to secure digital data and the frameworks they rely on to influence communication. The goal of cyber-security is to always protect sensitive information from the hands of vindictive third parties like business competitors, resentful employees, hackers, politicians, rivals or enemies. The Internet works like any other remote system, where telephone number corresponds to a particular device, a home or building address corresponds to a particular geographic location. The interconnected universe of computers and mobile devices frames the Internet which offers new difficulties for countries on the grounds that territorial or national outskirts don't control the progression of data that can be managed. The Internet of Things, cloud computing, smart applications and big data are currently working in a domain of regulations which incorporate a complex security scene. Privacy, on the other hand, refers to consumer's personal information and their capacity to completely know their rights with regards to how information about them is gathered, utilized and shared. All consumers' information must be kept safe, and its uses disclosed to the buyer in clear terms and basic language with the aim of transparency. Digitization has exposed businesses to advanced vulnerabilities, making cyber-security and privacy more significant than at any other time in human evolution. The world experiences cyber-attacks and privacy issues every now and then as the more technology advances, the more this attack persists. Every now and then there is always an update on another information break or digital security risk making the headline. From phishing, spear phishing and whaling, ransomware and malware, to ghost ware, blast ware and DDoS, dangers are becoming progressively popular, particularly in light of the fact that when an answer for one issue is discovered, another variant springs up. The general masses are the frail connection in the security chain. The human blunder is an immense reason for digital assaults and information breaks.

For making privacy strategies more reasonable and straightforward, legislative bodies have prescribed giving policy in a multi-layered arrangement. A short security notice on the top layer must offer people the center data required which incorporates at slightest the character of the specialist organization and the motivation behind information processing. Furthermore, a reasonable sign must be offered with reference to how the individual can get to alternate layers introducing the extra data required, for example, data on regardless of whether the individual is obliged to answer to the social media service provider's inquiries, and on the lawful privileges of the information subject on the social media sites should be clearly stated and well understood so that there is a level of control of personal information put up on social network sites. Privacy issues could affect the trust and confidentiality of websites, the more reason why strategies and plans should be put into action for a friendlier framework of all registered websites for security and data protection. Major issues of privacy concerns span across data protection measures taken by commercial websites to build trust. Also the fear of personal information being hacked and disclosed to the wrong party. Standard protection settings on most websites have individual information and utilization behavior stored, investigated, and transmitted to third parties so that the tastes of the users end up noticeably known to marketers that are permitted to target users with customized marketing strategies and continuous pop-up adverts whenever they are online, they keep on seeing this until they click on the adverts, close them from google, delete their browsing history, clear out their cache before third-party adverts actually stop; this is more like stalking them to make decisions on the spot or keep using particular products and services they know or patronize issues that have been raised on the impact of larger corporations like Facebook, Amazon, Instagram, YouTube, Alibaba, twitter and several others and their plans towards data protection of users who sign up and disclose personal information that is given to third parties who do not engage in privacy law or think about privacy concerns in their marketing strategies, but are primarily concerned with revenue generation. For instance, most social network sites generate a whole lot of money selling the personal information of its users without their consent or permission, thereby exposing them to certain threats or dangers. The World Economic Forum (WEF 2013), in their global dialogue on emerging issues surrounding data collection and use of personal data, identified the following areas to be considered in privacy concern issues in the future of digital era which are protection and security, accountability and rights, and also responsibilities for using personal data as critical areas to investigate the outcome of privacy laws on privacy concern issues.

The internet is a universe on its own, and every device is like one's home in this new universe. This home must be protected from Invaders. Security as a concept, be it digital or physical, must have certain attributes that make it feasible. Many companies have been hacked off the competitive space because of weak cyber securities. These even include crypto-currencies like Bitcoin.

For cybersecurity to be effective and impactful, it must take a comprehensive approach, where security is intrinsic and extrinsic on every part of the system from user to application and everything in between. This can be categorized under Policy or Processes, Technology, and People.

**Policy**: — A policy is a documented blueprint that goes along the beliefs and the goal of a company. A security policy therefore, is a documented measure to ensuring the security of the company putting into perspective the broad theme of such a company. A proper security policy states what the borderline of that company is and the punishment for breaching it. Security policies must be comprehensive. On a more personal level, security policy deals with the stated measures to ensure the safety and security of one's networks. The importance of security policies cannot be over emphasized. Security policies among other things serve as a forewarning to potential violators from both technological and legal frameworks. The organization must also be able to audit its policies and enforce them likewise. This is a cogent part of what a good security system must incorporate on the cyber space.

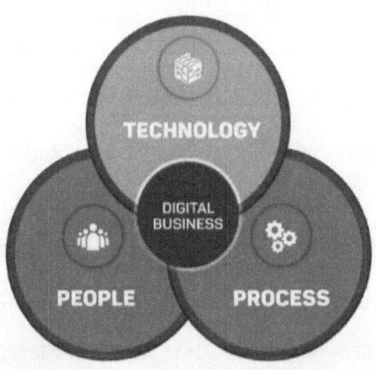

**Technology**: — To ensure the safety of a company or an individual on the internet, there are existing technologies that can be used to bring about the desired safety. Part of these technologies include:

•        Firewalls: This helps the individual or company to protect the traffic on its networks. It is a great tool to reinforce the network's security and also act as a spy dog against invasion.

•        Google Drive — Cyber-security in its broadest definition is summarized as a way in which a company or an individual ensures that information is not lost or stolen. Google drive provides an internet powered storage for the secrets of organizations which can be another security measure.

•       Secure WiFi networks — WiFi might just be the greatest tool to hackers and cyber bullies. The company whose security is taken seriously must secure its WiFi networks against the Invaders. WiFi automatically connects them with the network and gives the express permission into documents and information.

•       Encryption Software — This perhaps is the ultimate security need of every company and individual. This software distorts the stored information into unreadable codes and patterns which the Invaders might not find ways of deciphering. The software is of utmost importance to every industry for secure storage of information.

•       Regular backup —The information of a company must be backed up regularly and inspected to remove suspicious links and networks.

**People**: — Apart from the available technologies and policies, the employees of a company must also ensure the security of such company. The security of each device must be the duty of the handler while the overall security of the company's networks must be the duty of every employee.

The comprehensive security of a company must be end-to-end to bring about efficiency of management.

To create an impactful end to end security system, we need to understand the anatomy of the cybersecurity attack. We will take two examples: The Cyber Kill Chain by Lockheed Martin, and Full attack continuum by Cisco Systems.

**The Cyber Kill Chain**: — Lockheed Martin developed this as a means of identifying and preventing cyber intrusions. The cyber kill chain is a stage by stage analysis of cyber invasion pertaining to each network. The cyber kill chain observes the structure of invasion and the launch of an attack. The invasion is in seven stages which are identified as:

- Reconnaissance
- Weaponization
- Delivery
- Exploitation
- Installation
- Command and control
- Actions on objectives

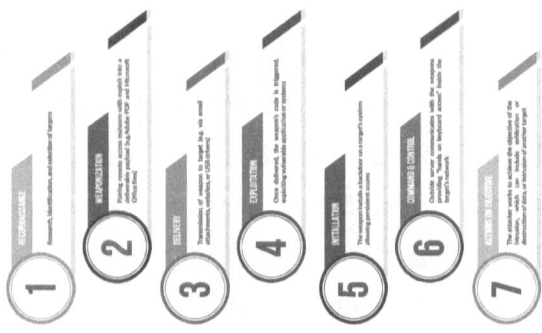

According to Cisco Systems Cybersecurity Strategy, it is viewed as three different phases, Before, During, and After. It is referred to, as the cybersecurity attack continuum. See figure below.

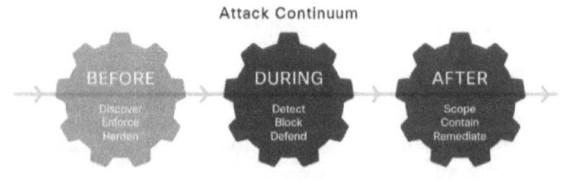

The cybersecurity architecture must provide comprehensive solutions and strategies that address every phase from all aspects: Policy, Technology, and People. See figure below:

| Firewall | VPN | NGIPS | Advanced Malware Protection |
| NGFW | UTM | Email Security | Network Behaviour Analysis |
| NAC & Identity Services | | Web Security | Adv. Malware Sandboxing |

Defense in depth and layered security is the name of the game, breaking it into pieces and defending against attacks one piece at a time. Relying on machine learning and automation to help security departments in each organization predict, detect, and defend "LET THE MACHINES PROTECT MACHINES". Map best of bread solutions to each phase of the attack, connecting the dots and solving the cybersecurity puzzle. See figure below:

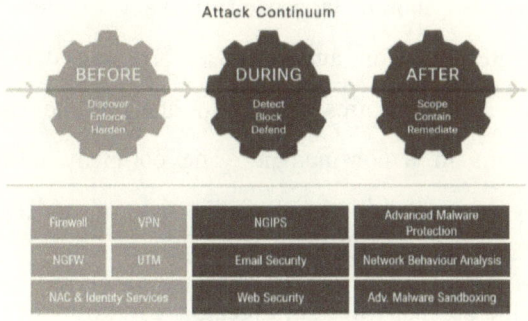

Cisco's approach to cybersecurity is considered to be one of the most advanced and impactful method; it is defense in-depth with layered approach. For added security, the cyber kill chain should be incorporated into the company's safety, ensuring quick identification of the intruder before an attack.

**Security Operations Center (SOC):** — A central security unit must be set up in a company to adequately monitor security issues from one network to the other. The sites and data of the company are supervised from this center and this makes the efficiency a bit more comprehensive.

## PRIVACY REQUIREMENTS

Companies should approach privacy issues with deliberate precision and purpose. Keeping data in a private space requires efforts to analyze the risk of operations in a confinement. The company needs to clearly have a full view of the risk of undertaking privacy in a digitalized world.

Also, the company needs to create a carefully done data policy as well as implement the tools for data protection.

The resulting effect is therefore a meticulous management of privacy and nabbing of offenders on legal grounds in the data policy.

GDPR is a head start plane for European companies on data management and breach of stated policies. While in the states, there are also regulations on privacy policy. However, the companies must also enforce their inner policies.

## PRIVACY RIGHTS AND CYBERSECURITY

Security enthusiasm for cyber-security includes building up protocols and successful oversight with respect to when, why and how government organizations may access individual data that is gathered, held, utilized or shared. Organizations and government establishment in the United States of America share obligation regarding the safety of customers' online individual data. There is no single government standard for information security that implements fair information practices (FIPs), to manage and authorize buyer security rights with respect to information accumulation, maintenance, use and sharing of individual data. The government approach has concentrated not on the security of individual data, yet on the motivation behind the data gathering. The historical backdrop of U.S. government organizations leading authorized and unsanctioned observation of local correspondence by conspiring with broadcast communications and wire correspondence organizations is notable. Domestic surveillance initially started as a method for obtaining data on crimes and immediately moved to record individuals' commitment to social or political exercises, their activities, and get togethers. The government likes to put control on the society to keep it in checks and balance; they spy but not all the time. The National Security Agency is no novice at diving into individual information that is verified by law or structure from snooping.

## SOCIAL NETWORKS

Social network platform connects people from various parts of the world through the use of the internet. Social network platforms are widely used by individuals, organizations and government agencies to communicate and interact with people through online communities. Social network platforms allow users to create profiles for free where they can share personal information, contents, pictures, beliefs and philosophies, and interact with other users whenever they are online. Business models represent ways a business can make money from its users, products, services and various offers. Social network platforms are allowed to generate revenues and make profits ethically, and must do their best to protect users' privacy. Despite the awareness of their private information being disclosed and sold to the third party for revenue generation, there is still the need for social network business models to have data protection measures for users in order to build trust and confidentiality

## BUYER CYBERSECURITY INTEREST

Online buyers have been misled by digital dangers as spyware, viruses, malware, fraud, cyberbullies and other forms of misrepresentation that mislead people to subscribe or invest into fake schemes. Online customers routinely succumb to wholesale fraud, just as spam, phishing or pharming assaults. Purchasers are confronting trust issues of identifying which product or organization protects them from all forms of cyber-attacks.

## CYBERSECURITY IN POLITICAL ADVOCACY

For people and associations that depend on the Internet for research, political advocacy, access to data, political cooperation, raising money, promotion, conflict resolution, political discourse, the spread of data or voting demographics, the fact remains that cybersecurity is of great importance for government, academics, individuals and organizations. Cyber-attacks restrict political activities, for instance, deceptive campaigns and propaganda that can, one way or another, compromise electoral activities and voter support by sending messages that are tricky or deluding with respect to the guidelines for voter cooperation on decision day. Digital assaults deny advocates access to the Internet as well as communication by denying them the best way to participate in a wide scope of political and online exercises that could incorporate security and unity, or enable electorates to know and comprehend approach that impacts their lives.

## BUSINESS CYBERSECURITY INTEREST

Today several organizations, both small and large, are at risk of cyber security which is within and outside their environment, some in which they have no control over their data and company secrets. Critical systems in organizations could be tampered with, ranging from information ruptures, burglary of organization, spying, assaults on PC systems and harm to basic frameworks. Due to the risk of cybersecurity, most organizations are using high tech solutions like cloud computing to secure their data and verify information from third parties. Nonetheless, distributed computing has colossal security and protection dangers identifying with reliance on dishonest or unevaluated outsiders. Businesses that have integrated digitization into their business model must always be updated to business applications that minimize cybersecurity of their organization, stakeholders, staff and customers.

## CYBERSECURITY IN NATIONAL SECURITY

The digital dangers to any country can go from interruption of an organization's systems or data services to people in general based on the kind of cyber assault and its reach. National security can be at great risk when the government is not a step ahead of cyber-attacks. There are negative reviews that come to countries that are constantly attacked in their cyberspace, such as genuine harm to framework undermining life or property. Digital assaults or occurrences that undermine the order and control structure of the national government or its advantages, including national barrier, crisis reaction and financial frameworks are of developing concern. The digital foundation of a country must be treated as a key national resource. The new mission is to dissuade, identify and shield against disturbances and assaults in the most effective way.

## POLICY

The internet is worldwide, yet the opportunities that are secured by sacred rights, human rights standards and legitimate organizations are characterized by settlement or based on region. Cybersecurity might be characterized by governments; however, it will affect numerous rights and common freedoms appreciated by free individuals all throughout the world who participate in digital communication. The opportunity of articulation, opportunity of affiliation, monetary chance and political talk might be re-examined to enact suitable policies for governments and their citizens. Choices about how to characterize cybersecurity and who will characterize it might influence Internet discourse, the opportunity of articulation and access to data. The individuals who have dealt with Network Neutrality comprehend what control of interchanges over the Internet may mean. Be that as it may, in the domain of government cybersecurity, openness and oversight probably won't be pieces of the procedure.

## AUTOMATION IN CYBERSECURITY

Current cyber-attacks have become vigorously automated, to the extent that organizations try to avoid cyber-attacks manually. The battle progresses towards becoming man versus machine, with organizations being at a higher chance of losing; it is necessary to battle fire with fire or machine with machine into cybersecurity activities. Automation makes everything fair, decreases the volume of dangers and considers quicker aversion of new and beforehand obscure threats. Easy access to information can put organizations and their clients at genuine hazard. A blend of AI and Automation can be utilized to counter these dangers and give understanding into dark and sinister movements on frameworks, systems and foundation. In Sanksshep Mahendra's report on AI + Automation, he analyzed the accompanying contextual investigation as a danger scene: In 2017, the average number of ruptured records by the nation was 24,089. The country with the most breaks every year was India with over 33k records; the US had 28.5k. (Ponemon Institute's 2017 Cost of Data Breach Study) The Equifax break cost the organization over $4 billion altogether. (Time Magazine) The normal expense in time of a malware assault is 50 days. (Accenture) Pretty much every business or organization is advanced now and manages necessary information and foundation for their activities. Digital dangers put a strain on finance, time, and other resources that could be put into good use for maximum returns to the organization. In spite of the mindfulness about these dangers, there is next to no speculation and preparation among organizations to make a preventive move and get a master to exhort till they are hit. Digital Framework on cybersecurity must begin with the system and a profound comprehension of the procedure, which will help in automation.

It is evident that protection of information, better known as data in the digital world of business, creates a secured digital environment for business to strive in this new world where data is the new currency. Security of process information and privacy of end-users improve the reliability and positive interaction of all parties involved, effecting the transformations experienced in this digitalized new world. Now, let us discuss what the world will look like in the nearest future, considering the rate and magnitude of revolution brought by technological transformations.

# CHAPTER 20

## A LOOK AT THE WORLD IN THREE DECADES

We have been looking at what the world is currently going through with respect to the technological revolution. Thanks to this same advancement in technology, we can now foresee what the nearest future holds for us in terms of technological revolution. It is evident that the technology jump in the next decades will be immense as we have seen in the previous few decades. Some components of our reality will change to the point of being indistinguishable, while others will remain recognizable. The change we have seen from way back from 1995-2015 was just the beginning of the internet, we worked in desk areas, and our PCs were large and supported by Windows 95. There were no touch screen telephones or LED screen TVs. People had to use their VCR to watch home videos. Things being what they are, what will our reality truly resemble quite a while from now? What does the future hold for the food we eat, the innovation we use, and the homes we live in? It is enticing to reveal the basics of food pills, flying cars, or living on the moon; however, the truth will presumably be less energizing. The world in a couple of years will most likely be very similar to how it is today, yet more intelligent and increasingly programmed. A few developments we probably won't see, while others will prompt rapidly, changing our lives for eternity. The impact of technology on every aspect of our lives that we have seen is only the beginning, we are yet to witness the highest growth in seismic disruptions, fueled by the continuous rise of computer power, at a much lower cost, with higher connectivity, and the fabric of society will reach 5B and beyond. Customer expectations will be at all-time highs. Things will be ordered, and instantly delivered, the industry giants will compete for the Data, and through data, and everything will become autonomous. AI will impact every function just as humans do; more than 100 Million jobs will transform, change and potentially vanish. The jobs of the future will look absolutely nothing like today's jobs. Education will transform completely. For the new skills we will have robots exploring planets and potentially building a life there for us. We will have some living humans on Mars. We will have robots exploring the ocean/sea for our benefits. We will have robots working in farms for us, robot friends. Let us unleash the imagination... We need to add futuristic thinking to most industries. That is all the data about the Impacts of Future Technology in The World. As time passes, innovation will develop significantly. Along these lines, it will make a splendid future for individuals around the globe. At that point, would you say you are prepared for a superior world?

## THE FUTURE OF FOOD

The following real food insurgency is found in vertical farming in which we develop food in AI-controlled vertical structures as opposed to flat land: hydroponic plants for foods grown from the ground and in-vitro cloned meat. This change is, as of now, occurring. We are seeing food scientists developing kale, spinach and different greens under LED lights in a plastic plant. Vertical cultivating, hereditarily changed (GM) harvests and engineered meat will be reactions to the developing requirement for more prominent food effectiveness as a population keeps on developing. Yet, there will likewise be a hesitant acknowledgment that we, as a whole, need to have an eating routine, one that is more plant-based and less dependent with respect to lab grown or artificial nourishment. Some could be eating bugs in 2035; officially prevalent in parts of Asia, creepy crawlies are protein-rich, low in fat and a decent wellspring of calcium.

Our bodies will be a lot more intelligent in the sense that they might become able to tell us exactly what type of nutrients we need and what we should eat to get it in the most efficient way possible. Of course, this won't be possible without the use of technology or digitalization.

The use of robotics in the food sector might also help us get more out of the 76% of earth's water reserves and thus we'd have more options and room to explore our diets.

As farming is becoming a lot more intelligent with digitization, I imagine that it might reach a period where there would be plants that can be grown in as quick as a day, eliminating the need to overwork the farms in the switch to a plant-based diet; even beyond that, there will be farming and planting on demand (call it: Seed as you Need). The possibilities are limitless, when we combine multiple technologies that allow us to build the intelligent farm that is monitored and watered by drones. We can create unlimited food supplies in small farms, flat farms and vertical farms.

With more than 71 percent of earth being water, of which 96 percent is ocean, can robots help us? Perhaps, the majority of our protein diet in lieu of insects and crawlies will be sourced directly from the seas and aquatic bodies.

## THE EVENTUAL FATE OF RELATIONSHIPS

Not at all like the films showing going gaga for an artificial intelligence (AI) working framework that has feelings, the web has changed the manner in which individuals meet and fall in love. Internet-based dating and area-based administrations, for example, Vine, Snapchat, and Grindr have opened up conceivable outcomes that enable individuals to look past their quick friends, former partners and colleagues. Where marriage used to be the start of an organization, it's turning into the finale. We are winding up the old social standards. This will affect the connections we structure, with fewer individuals picking conventional marriages and more individuals staying single for longer, if not for eternity. Singles are becoming more acquainted with a partner before they get married, where marriage used to be the start of a relationship.

Will technology give us the means and be the platform to address this issue, could it play the role of intermediary to overcome the social structure that has been disrupted and is on the verge of being completely broken? I believe it could and would; let us see what will happen in the next few decades.

### The Eventual Fate of Work

When it comes to work and jobs, this is a rather very sensitive topic, that there are many views on. One view is the world is collapsing, where humanity will be a competition and a threat to robots; while another view is there will be no impact on the future of work, and for every job that is gone, it will be replaced with another. I believe the future will be somewhere in between.

Rather than people working with machines, technology is probably going to make a few employments repetitive: cabbies replaced without anyone else driving Uber vehicles, receptionists replaced by robots, specialists outmaneuvered by calculations that can connect to tremendous therapeutic databases. Clearly, there will be new occupations made: driving Uber taxis, software engineers, genome mappers and bioengineers, space visit aides and vertical ranchers. Innovation will keep on disrupting organizations and take out occupations, making new callings we cannot imagine. We will see an arrival to progressively dynamic nearby networks as individuals work remotely from their homes.

As the future is approaching us at maximum speed, we need to do something about this. I believe we can definitely design the future of work, discussed in chapter 21, designing the extraordinary future.

## The Future of Health

Technology would become a greater part of our everyday life as a result of its significant impact on our life. It is estimated that people spend 40% of their earnings on their physical and emotional wellbeing. Hospitals happen to be one of the most expensive systems today, but in the future, it would be a different ball game as healthcare practices would be very affordable and seamless. In the nearest, future, prevention would be the trending concept in healthcare as opposed to medication. Future health services will attempt to keep individuals out of them. Aversion will turn into the concentration as we deal with our wellbeing data, utilizing self-observing biosensors and savvy watches to assemble wellness information ceaselessly; web applications will crunch the information, synchronizing to electronic wellbeing records. Utilizing these numbers, health organizations will be able to use the data collated to develop a model or services for general wellbeing that can foresee future issues. Being cautioned, patients will almost certainly make a move early by changing their lifestyles or taking medications that are custom fitted to their individual DNA. Medical practitioners would be able to leverage on technology to practice online through telehealth platforms. There would be in-home or office monitoring of patients who desire to consult their doctor. Patients in remote towns would be able to get specialist care from any country they desire. Telehealth platforms will make in-home patient observing the standard for those who need it. Genome mapping will prompt customized medications and 3D-printed substitution organs. In the meantime, the unmanned aerial vehicle (UAV) innovation will be utilized as driverless rescue ambulance drones. More attention would be placed on keeping fit and living healthy.

This has led to innovations of people empowering organizations to map clean drinking water projects with mobile technology. There are many social initiatives by leaders in the industry as part of their responsibility to use technology and innovation for the greater good. As an example, Cisco supported Water for People in 2010, and made further investments to Akvo in 2012, for the creation and development of the FLOW application. This mobile tool collects and analyzes data on water projects in the developing world and has benefited more than 3.5 million people.


CLEAN ENERGY

More emphasis would be made on a cleaner energy source as opposed to fossil fuel, which has driven our earth into the edge of catastrophic results such as global warming. Technology would emerge to take care of global issues that have resulted from the use of fossil fuel; the world would be introduced to other energy alternatives like solar energy, wind turbines and other forms of renewable energy that are environmentally friendly. Technology innovation in solar energy would drive down the prices of solar cells, which would make it easier for mass adoption of clean energy sources just as the era of mass computer adoption based on Moore's law said the lower and faster the innovation, the more and higher the rate of adoption. Clean energy would become more affordable, thereby leading to mass adoption.

For example, Renewable sources of energy accounted for 40 percent of German electricity production in 2018, up from 38.2 percent in 2017 and 19.1 percent in 2010, according to REUTERS, published January 3rd, 2019. The cost of solar energy product keeps going down, while the efficiency, optimization, and ability to store keeps going up, exponential growth in both direction, this makes a future with abundant energy at low or no cost; that is the potential future I see and predict.

## INFRASTRUCTURE

We will be able use materials that are easily recycled or reused, while robots and 3d printers will become capable of building what we need in hours or minutes, right before our eyes. Not only would we be able to travel directly through ocean, we would also have the means and method with which to live in in them while traveling, like an underwater cruise of sorts. Perhaps the need to even take time to evaluate property might become redundant altogether if the world manages to link information about everything to one source, making it easy to choose a house and have it built in minutes.

## ACCESS TO EDUCATION

Not everyone on planet earth has access to quality education today. The internet has made education and personal development available at our fingertips. Education on basic and advanced topics can be learned from the internet and mobile applications built for learning. The future of education would have educational institutions from all over the world, recording their lectures and publishing educational materials on the web for everybody who needs to watch and learn. The more internet connectivity improves, the better it becomes for a larger population to access education without necessarily being in a school environment. A ton of current jobs on the planet is probably not going to exist in the closest future. The world is changing quicker than we might suspect and the manner in which we learn things and what we realize ought to be changed. Bosses ought to make a drawing in learning society at work by enabling their staff to take responsibility for expert learning and development.

## MORE EXPECTATION FROM THE FUTURE OF TECHNOLOGY

Nanobots will connect our cerebrums straight to the cloud. By 2050, nanobots will connect our minds directly to the cloud; it will give us full virtual reality experience from inside the sensory system. Much the same as we do now with our smartphones, we will almost certainly do with our brains, we'll have the option to grow our neocortex in the cloud. Also, disregard memory issues, trust issues, and several issues we are experiencing because we would be able to wipe out certain issues from our memory.

Artificial intelligence will turn into a positive employment helper. A whole lot of people stress over AI in our lives as they feel that toward the end, robots will displace humans, and we won't have jobs that require humans again. According to Forbes, in 2020, AI will turn into a positive employment helper, making 2.3M employments while disposing of just 1.8M jobs. Furthermore, we are discussing 2020, just in a year, so how about we see what openings it can acquire for us in the space of 30 years. IoT innovation will change the fast-moving consumer product designs. By 2020, IoT technology will be in 95% of electronics and mobile gadgets for new product designs that would be in line with the internet of things framework. Also, by 2050, it is relied upon to have everything connected with the cloud in the internet.

The travel industry would experience space tourism. Space travel could be doable in 2050, for the extremely rich. Rocket organizations like Jeff Bezo's Blue Origin and Elon Musk's SpaceX will push the trend for space travel tourism, as this will be practical in the decades to come.

Self-driving cars will make driving more secure. This region of AI could significantly decrease accidents and wounds on our streets, despite the negative headliners about self-driving cars, something good is going to happen in this field soon. As indicated by a report by Stanford University, self-driving cars would not only decrease traffic and accidents but give us more time for ourselves and our loved ones. Self-driving cars and shared transportation may influence where individuals live.

In the next 3 decades to come, the world would face shortages of IT specialists. There is an immense interest for engineers in Europe, yet most IT instruction is exclusively accessible in English. With 24 official dialects spoken in Europe alone, it's nothing unexpected that non-local English speakers like to learn in their own language. Anyway, you can begin seeing many coding foundations, schools and colleges concentrated uniquely on preparing junior designers for explicit organizations' interest, being it a language, training and so forth. Despite the fact that this issue won't be understood by 2020, with these new IT colleges and institutes, we can fix the circumstance by 2050.

Drones would reach places where humans have never been before and create maps of such areas. Somewhere down in underground mines, a few zones are distant. In any case, an organization like Inkonova began to take a shot at creating drones that fly, drive and climb and use laser innovation to access any zone and make a 3D map of them. With this propelling technology innovation, we will most likely push human reach to any space unreached or inhabited by humans.

With better healthcare systems, the world population will grow to 10 billion humans or more.

## THE EVENTUAL FATE OF TECHNOLOGY

What it could resemble: Technology supports all that we've taken a look at up until now such as food, wellbeing, jobs, tourism, healthcare, energy, education and work. We're going into a future where improved battery innovation will empower better electric vehicles, individual flying machines, Hyperloop transportation, private space, the travel industry and automation services. We'll wear Band Aid-style wellness sensors on our skin, charge our gadgets utilizing remote power, let algorithms upgrade and secure our homes and have menial helpers (the up and coming age of Google Now, Siri and Cortana) to enable us to deal with the surge of information and understanding them.

But whatever happens next, it will be a great time to be alive. A portion of this may occur. Or on the other hand none of it. Three things, be that as it may, are sure: Technology gadgets will get littler, more intelligent and less expensive. Truth be told, it will get so little, smart and modest that we'll have the option to put computers and sensors into nearly anything. Our refrigerators will disclose to us when we've come up short on milk, waste bins will be able to tell when they are full, TVs will see when we've quit watching and turn themselves off to save power. We're making progress toward the internet of things where everything is associated, not exclusively to the internet but to each other. What's to come is… flighty!!! Foreseeing what's to come is famously dangerous, particularly when it does not become feasible. In 1883 Lord Kelvin, leader of Britain's Royal Society, pronounced 'X-rays will demonstrate to be a lie.' Arthur Summerfield, the US Postmaster General in 1959, anticipated that mail would be delivered in few hours from New York to Australia by guided rockets.

What's more, we ought to be happy that Alex Lewyt's 1955 idea of 'atomic controlled vacuum cleaners' never made it to the planning phase. Be that as it may, whatever occurs straightaway, it will be an extraordinary time to be alive in a world of abundance.

# CHAPTER 21

## DESIGNING THE EXTRAORDINARY
## FUTURE

While there are too many predictions about the future, some are good, others are great and some are really bad, I fundamentally believe, we have a unique and an unpresented opportunity to invent and design our future. The extraordinary future that is powered and enabled by technology.

The impacts of advances in technology are many and varied. The increase in processing power, as defined by Moore's Law, brought advanced computing power down from room-size installation to the size of an A4 pad. Cost spiraled down similarly to within reach of millions of people. Storage capacities increased as hardware size decreased, and screen resolutions sharpened to almost analog levels. All these advances are accessible not just in size and cost, but also in ease of use. The reliability of hardware and software plus advances in design and usability broadened usage even more. So now, a large percentage of people in the world have a computer in their pocket in the form of a smartphone, broadband usage is through the roof and most major services, like government, utilities, banking, etc., are delivered online. The process by which access to technology rapidly continues to become more accessible to more people. We can see how technology has democratized everything when things become digital, they effectively become free. We can take many examples in every sector and see how digitization drove the cost to zero or near zero. The example of genome sequencing cost, according to the National Human Genome Research Institute (NHGRI); (https://www.genome.gov/about-genomics/fact-sheets/DNA-Sequencing-Costs-Data); in 2001 cost one hundred million dollars, and now it is under 100 dollars, see figures below:

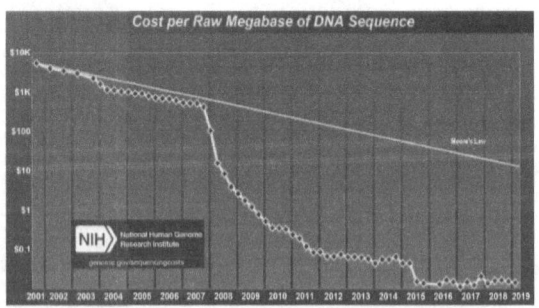

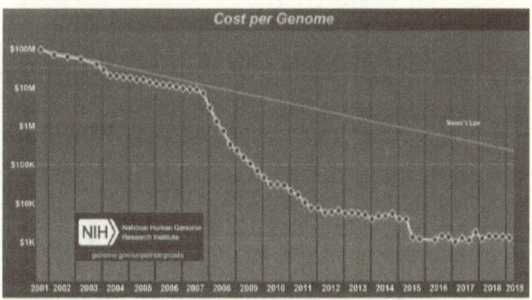

New technologies and improved user experiences have empowered those outside of the technology industry to access and use technological products and services. At an increasing scale, consumers have greater access to use and purchase technologically sophisticated products, as well as to participate meaningfully in the development of these products. Industry innovation and user demand have been associated with more affordable, user-friendly products. This is an ongoing process, beginning with the development of mass production and increasing dramatically as digitization became commonplace.

So many studies argued that the era of globalization had been characterized by the democratization of technology, the democratization of finance and the democratization of information. Technology has been critical in the latter two processes, facilitating the rapid expansion of access to specialized knowledge and tools, as well as changing the way that people view and demand such access. Scholars and social critics often cite the invention of the printing press as a major invention that changed the course of history. The force of the printing press rested not in its impact on the printing industry or inventors, but on its ability to transmit information to a broader public by way of mass production. This event is so widely recognized because of its social impact – as a democratizing force.

The printing press is often seen as the historical counterpart to the Internet. After the development of the Internet in 1969, its use remained limited to communications between scientists and within government, although the use of email and boards gained popularity among those with access. It did not become a popular means of communication until the 1990s. In 1993 the US federal government opened the Internet to commerce, and the creation of HTML formed the basis for universal accessibility.

The Internet has played a critical role in modern life as a typical feature of most Western households and has been key in the democratization of knowledge. It not only constitutes arguably the most critical innovation in this trend thus far, it has also allowed users to gain knowledge of and access to other technologies. Users can learn of new developments more quickly, and purchase high-tech products otherwise only actively marketed to recognized experts. Some have argued that cloud computing is having a major effect by allowing users greater access through mobility and pay-as-you-use capacity.

Social media has also empowered and emboldened users to become contributors and critics of technological developments. The open-source model allows users to participate directly in the development of software, rather than indirect participation, through contributing opinions. By being shaped by the user, development is directly responsive to user demand and can be obtained for free or at a low cost. In a comparable trend, Arduino and littleBits have made electronics more accessible to users of all backgrounds and ages. The development of 3D printers has the potential to democratize production increasingly.

CULTURAL IMPACT

This trend is linked to the spread of knowledge of and ability to perform high-tech tasks, challenging previous conceptions of expertise. Widespread access to technology, including lower costs, was critical to the transition to the new economy. Similarly, the democratization of technology was also fueled by this economic transition, which produced demands for technological innovation and optimism in technology-driven progress.

Since the 1980s, a spreading constructivist conception of technology has emphasized that the social and technical domains are critically intertwined. Scholars have argued that technology is non-neutral, defined contextually, and locally by a certain relationship with society. Andrew Feenberg, a central thinker in the philosophy of technology, argued that democratizing technology means expanding the technological design to include alternative interests and values. When successful in doing so, this can be a tool for increasing inclusiveness. This also suggests an important participatory role for consumers if the technology is to be truly democratic. Feenberg asserts that this must be achieved by consumer intervention in a liberated design process.

Improved access to specialized knowledge and tools has been associated with an increase in the "do it yourself" (DIY) trend. This has also been associated with consumerization, whereby personal or privately-owned devices and software are also used for business purposes. Some have argued that this is linked to reduced dependence on traditional information technology departments.

Astra Taylor, the author of the book The People's Platform: Taking Back Power and Culture in the Digital Age, argues, "The promotion of Internet-enabled amateurism is a lazy substitute for real equality of opportunity."

## INDUSTRY IMPACT

In some ways, the democratization of technology has strengthened this industry. Markets have broadened and diversified. Consumer feedback and input is available at a very low or no cost.

However, related industries are experiencing decreased demand for qualified professionals as consumers are able to fill more of their demands themselves. Users of a range of types and statuses have access to increasingly similar technology. Because of the decreased costs and expertise necessary to use products and software, professionals (e.g., in the audio industry) may experience loss of work. In some cases, technology is accessible but sufficiently complex that most users without specialized training are able to operate it without necessarily understanding how it works. Additionally, the process of consumerization has led to an influx in the number of devices in businesses and accessing private networks that IT departments cannot control or access. While this can lead to lowered operating costs and increased innovation, it is also associated with security concerns that most businesses are unable to address at the pace of the spread of technology.

## POLITICAL IMPACT

At a demonstration, banners read, "Democracy needs anonymity – stop data retention" (left) and "Liberty dies with security" (right). Some scholars have argued that technological change will bring about the third wave of democracy. The Internet has been recognized for its role in promoting increased citizen advocacy and government transparency. Jesse Chen, a leading thinker in democratic engagement technologies, distinguishes the democratizing effects of technology from democracy itself. Chen has argued that, while the Internet may have democratizing effects, the Internet alone cannot deliver democracy at all levels of society unless technologies are purposely designed for the nuances of democracy, specifically the engagement of large groups of people in between elections in and beyond government.

The spread of the Internet and other forms of technology has led to increased global connectivity. Many scholars believe that it has been associated in the developing world not only with increased Western influence, but also with the spread of democracy through increased communication, efficiency and access to information. Scholars have drawn associations between the level of technological connectedness and democracy in many nations. Technology can enhance democracy in the developed world as well. In addition to increased communication and transparency, some electorates have implemented online voting to accommodate an increased number of citizens. Abundant connectivity, smart phones and mobile apps brought people and governments much closer, and created a platform of communications and engagement. This created an unprecedented opportunity to enhance the quality of life, and respond to public demand efficiently and at speed.

Technology brought about abundance. Technological innovation is great for consumers. As technology gets more advanced, prices drop, and products get better.

## TECHNOLOGY MAKES EVERYTHING AFFORDABLE OR FREE

Very few luxuries would exist without technology to create them, but that doesn't prevent technology from being the thing that eliminates the luxury status as well. Just think about how much more you can do with your modern smartphone than Gordon Gekko could ever accomplish with his expensive Carphone in the 1980s. Medical care today is orders of magnitude better and safer than it was just a century ago. The richest kings and queens could have spent all their wealth, and they would never have had access to the basic vaccines, antibiotics, and treatments that we have developed in recent decades.

Technology is the biggest threat to luxury because of how it is developed and commercialized. There's a virtuous cycle between mass market devices producing the biggest profit and that profit being reinvested in the development of further mass market devices. Almost no one is investing billions into researching and developing a thing that would only ever be accessible to a limited, extremely wealthy clientèle. Even Bugatti, the car company that defines opulence and excess, is developing technologies and manufacturing methods that will later trickle down into use in more attainable models from its parent company Volkswagen. Cars of Bugatti's ilk never go down in price, but the innovations that they pioneer often do.

It's no secret that technology is threatening to take away jobs. For all the talk about robots working alongside humans rather than replacing them altogether, automation's higher efficiency, lower costs, and increasing capability mean eventually workers will be removed from the equation in many jobs. No one wants to be replaced by a machine, but there's a silver lining.

The counterbalance to technological unemployment, is the demonetization of living—in other words, pretty much everything we need and do in our day-to-day lives is becoming radically cheaper, if not free, and technology's making it happen. The most obvious and tangible example of this phenomenon is, of course, the smartphone. Twenty years ago, we had a bunch of different things that each performed a single function: a camera took pictures, a flashlight lit up the dark, a TV was for watching shows, a VCR played movies, a boom box played music, and so on and so forth.

Now we have all that and more in the palm of our hands. More significantly, though, we got most of it for far less than in the past. If you added up the cost of all that hardware 20 years ago, you're looking at thousands of dollars—now reduced to a few hundred. Similarly, the average smartphone is for $50 and in developing nations holds millions of dollars' worth of software.

Demonetization is a notable technological disruption, happening after digitization but before democratization. Taking money out of the equation for a given product or service is a key part of making that product or service available to everyone. Below are just a few of the examples of demonetization across various industries.

## COMMUNICATIONS

If you don't have a smartphone or computer, you can't have your data collected—and companies want your data. They want it so badly, for which I predict in the very near future some companies will start giving smartphones for free, specifically in the areas of the world where the vast majority of would-be consumers aren't online yet. The data they will be collecting is way more valuable than the device they will give away. Just imagine, if they invest $50 in a device, and a user uses it for a year, can you estimate the value of that data?

According to statista,

(https://www.statista.com/statistics/330695/number-of-smartphone-users-worldwide/)
more than 3.3 Billion people use smartphones globally, as of July 26th, 2019, with an expectation to grow by 500 Millions over the next couple of years.

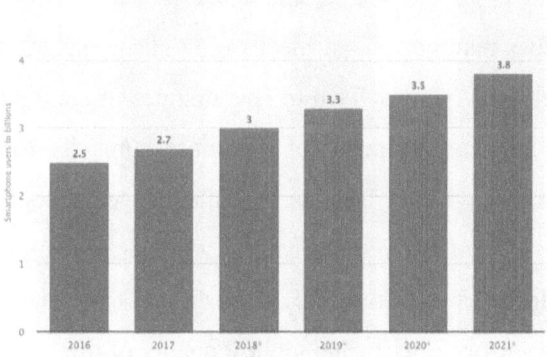

Those will join the global fabric of society, they will benefit and they will contribute to the global economy and communication.

## ENTERTAINMENT

We used to drive to Blockbuster and pay a few dollars to rent one movie. Now we can pay a low flat rate and watch as many movies and shows as we want each month. Or we can watch stuff for free; YouTube streams millions of hours of free video per day.

## ENERGY

The poorest countries in the world are the sunniest countries in the world, and solar power is becoming cheaper than coal. That means ultra-cheap electricity in developing nations. Two to five minutes of sunlight can fulfill the energy needs of the entire humanity for one full year; will we be able to build the technologies to harness it? I believe we can, if we address the energy challenge, we can create a world of abundance that allows for prosperity and highest quality of life.

## TRANSPORTATION

When you own a car, you have to pay for fuel, parking, insurance, tolls and maintenance—not to mention buying the car itself. On-demand ride apps like Lyft and Uber are changing the way people get around and are making it cheaper for them to do so. Why pay all that money for your own car when there's a service to get you from point A to point B at a fraction of the cost? Electric autonomous cars will disrupt transportation even more. Even beyond, imagine autonomous vehicles that are powered by solar. This is not only possible, it is available today, and I believe it will be available at a wide scale in the next decade, and will be much more affordable than one thinks; it could potentially be free offered by government or businesses as part of the services delivery.

## HOUSING

Self-driving cars will change the housing market by enabling people to commute from farther away more easily. The housing itself will get cheaper thanks to large-scale 3D printing. The 3D printing is advancing so rapidly and exponentially; reducing the cost to build, enhancing the quality, using recycled material, while increasing the speed to complete it. A very good small home can be built with a cost of under 4,000 US Dollars.

## EDUCATION

Students of all levels and ages can benefit from the new learning methods that the gig economy provides, giving them more choice of where and how they learn. Educated, technology-savvy graduates are choosing to work gigs instead of pursuing salaried, long-term employment. For example, about six million students have participated in the Cisco Networking Academy program since 1997. Regardless of their socio-economic background or gender, these students develop the expertise to master, succeed and lead in the digital economy. Together with their 9,500-plus educational partners in more than 170 countries, Cisco is building intellectual capital in networking, security and IOT technologies – the critical technical skills required by nearly every business on the planet. Unlimited educational material is available online. Over the past two decades, not only the educational material has been democratized, it has improved with quality, and reduced in cost to be offered at no cost. The educational platforms are available in all types of media, and in all languages, the democratization continues in an exponential speed.

## HEALTHCARE

Life expectancy has increased rapidly over the past century. Estimates suggest that in a pre-modern, poor world, life expectancy was around 30 years in all regions of the world. In the early 19th century, life expectancy started to increase in the early industrialized countries while it stayed low in the rest of the world. This led to very high inequality in how health was distributed across the world. Good health in rich countries and persistently bad health in those countries that remained poor. Over the last few decades, this global inequality decreased. Countries that, not long ago, were suffering from bad health are catching up rapidly. A good example is Living Goods' network of community health agents, who use mobile technology to improve the health of families in their communities. These entrepreneurs go door to door selling life-changing health and hygiene products and providing health education. Cisco began support in 2012 for the mobile technology tool that is now the backbone of this model. Living Goods has grown from 400 community health agents serving 200,000 clients in 2012 to 4000 serving three million clients today, including 675,000 children. Their efforts have lowered child mortality by 25 percent at an annual cost of less than $2 USD per person, while creating livelihoods for thousands of women. I have no doubt that technology will enable us to diagnose and treat every disease and illness; we are definitely progressing towards that, despite some opposing forces.

Since 1900 the global average life expectancy has more than doubled and is now approaching 70 years. No country in the world has a lower life expectancy than the countries with the highest life expectancy in 1800. Genome sequencing will transition healthcare from being reactive to proactive, keeping people from getting sick in the first place.

I view the world as rapidly demonetizing. A world where life's necessities are all cheap or free will be very different from the world we live in today. What will motivate people to work or be productive if they don't need money for the basics? What kinds of new innovations will spring up from people for whom these resources used to be cost-prohibitive? How will social constructs build around wealth and class shift?

These are all questions we'll need to contemplate as technology continues to demonetize our lives. As the old saying goes, the best things in life are free, and if this vision becomes a reality, we'll have to figure out which of the free things in life are best.

# CONCLUSION

In the broadest sense, from all we have been discussing in this book, technology has really revolutionized over the past decades with tremendous velocity and impact over the universe as a whole. Name it, from the first visit of man to the moon and now possibly to the commercialize tourism visits to other planets. Then we can definitely say that the earth is not the only planet impacted by technology. But we can say that earth has had a considerable fair share, from industrialization to commerce, internet, world-wide-web, automation, biotech, social media, the list is just endless. The predictable facts are that we haven't seen anything yet, we haven't attainted or gotten close enough to attain the full potentials of technology. Technology gives us the abilities to change the world: to cut, shape and put together materials, to move things from one place to another, to reach farther with our hands, voices and senses. We use technology to try to change the world to suit us better. The changes may relate to survival needs such as food, shelter, or defense, or they may relate to human aspirations such as knowledge, art or control. But the results of changing the world are often complicated and unpredictable. With an interesting level of comparing present technology with that of earlier times, as well as the technology in their everyday lives with that of other places in the world and the future, we can imagine what life would be like without certain technology, as well as what new technology the future might hold.

Lots of predictions have been made, lots of models developed to help industries across several sectors of the world economy, but the actual fact is that the more we try to predict what is to come with technology, the more knowledge is required, and the source of knowledge to this context is endless, and this is all thanks to technology itself. One thing that is constant in life is change. The way things were done in the 17th century is certainly not the way they're done today. William Shakespeare wrote in a different kind of verbatim called the Old English, and today playwrights with just as much literary prowess write in much more contemporary and innovative ways that tremendously increase user experience. The particular rise of a new generation of people known as millennial is not a fad, but a sign that we are fast abandoning the older, more difficult ways of doing things. We have crypto-currencies replacing our current payment systems and robots doing more work than we could have ever imagined, reducing the stress and infections associated with human manual labor. With a mixed economy system taking over most nations of the world, the future is bright, and it is already shining. It is certainly a better one because forward march is a depiction of progress and development. It is all about technology, and the sooner we start to experience it, the better for all of us.

In early centuries, letters and messages were sent through birds such as crows and ravens. Post office companies then came into existence, and the world felt more at ease with sending messages. Fax came before the possibility of electronic messages came to life, and I'm sure everyone might think of what could be better and faster than sending letters and documents via emails. Today video conference calls are used for connecting people from different ends of the world to achieve common goals. With 3d printing coming to life, virtually everything you might want to need in life will be printable just from the comfort of wherever you find yourself.

Revolution and transformation are endless. From the use of Horses for transportation to the invention of automobiles and possibilities of flying cars, like from the use of steam engines to the jet engine and the possibilities of engines that work at the speed of bullets. Since the dawn of time, humans have always been limited by what they do not know, often resulting in trial and error, fire brigade approaches, and emergency overalls. Since the dawn of the same time, mankind has strived to contrive necessary tools and technology that will assist the race in the pursuit of individuals as well as collective goals. Significant changes in technology have more or less culminated in huge shifts in social stratifications, including how people contributed to the development of society and how they make a living for themselves. The Industrial Revolution led to so many changes that happened on a large scale, especially as it concerns socioeconomic structures and the kinds of jobs people took and earned and lived from.

As of now, advancements in technology, more commonly known as innovations, are in a rapid manner making previously impossible things possible, such as the automation of a larger portion of work that used to be carried out by human hands. This very much applies to blue-collar jobs, by means of new age concepts such as robotics and the much sort after the Internet of

Things, and white-collar jobs by means of the proliferated narrative of Artificial Intelligence (AI). The wide range within which these technologies can be applied has both resulted in the destruction of jobs and the creation of new ones. So, we have to reckon with the power of innovation, and how it is affecting the way we do things. Just about a century ago, a memory card was so large it had to be transported with the use of cranes, but today, we can have an almost infinite amount of memory space on small chips, which further defines Moore's law in reality. With the way these advancements have disrupted the normal way of doing things, bringing in more efficient and easy ways, then imagine a world where technology is the future of everything.

Given the tragedies that happen around the world - the plane crashes, earthquakes, deadly volcanic eruptions, onslaughts, tsunami destructions, wars and even the much told-about global warming - it is somewhat beautiful if we can live in an age where there is no fear of the unknown, as science will be able to help us know what is going to happen each day and the next. Having such power, it is easy to envisage a world devoid of chaos, madding mayhems and crazy confusions, all of which have somehow placed the world on the imbalance which it currently is today. With a myriad of opposing views, opinions and courses of action, it becomes rather impossible to predict the future of any of the systems or sectors mentioned. Probably we would have been able to prevent so many plane crashes and missing ships in the Bermuda triangle. These incidents claim lives and destroy millions of dollars in seconds.

It is pretty much vivid, as of now, that technology will save the world, and in many ways as such. As advancements keep popping in different sectors and in different forms such a Big Data, Augmented Reality, Virtual Reality, Artificial Intelligence and robotics, lifelines for what we today know as planet earth could just be in the many. Countless bottlenecks being faced in a diversity of sectors, it is innovation that will become the savior of mankind. Beyond the whipping up of tacky machines and the conservation of the immediate environment, technology, by means of accurate prognostics will help countries govern better, economists analyze better and religious fanatics serve Supreme better, and education become more impactful. If we know the best ways things can be done, then we are entirely guaranteed nothing short of utter impeccability.

Rather than guessing and second-guessing, we would act, think, and speak with a 101 percent level of assurance, without having to worry about costly mistakes and glitches in the result processes. It would be a time when we would all participate in what will have already been a perfect world, basking in the ideal place to exist as a person.

Well, for one, nothing is perfect. And, based on popular conviction, the world cannot be perfect either. But what is pure perfection? Is it something that has been actually experienced or just subjected to the "it has to be" kind of verbatim? With the loopholes all up and about, it is easy for one to conclude that no one really knows what perfection looks like. But there is one thing which stirs hope in that course every day, and it is the belief that things can get better. So, when there is that chance - which has been proven time and again - it means that technology can make the world a better place by helping us predict things to come. This is one of the reasons I am glad to be associated with a leading technology innovation organization such as Cisco. The level of commitment to advance the circular economy, not only through the technologies put on the market, but across every facet of our business. This means thinking differently about how we manage our operations, design and build products and reuse and recycle assets. As more and more cities chart their own roadmaps to becoming smarter and more circular, Cisco continues to bring together technologies, thought leadership, and a growing partner ecosystem to support the journey. In my current role, I have the great pleasure to work with Governments, Partners and customers where I participate and help in building digital transformation strategies that allow societies and organizations to unlock the power of technology and design the future we all aspire for.

I am hoping this book will serve as an encouragement and aspiration to public and private organizations to start their digital journeys. My word to you, for you to keep up with the pace at which technology is transforming and revolutionizing the world, embrace technology in the fullness of its potentials in everything you do, and for all of us to partner to make the world amazingly great for all of us for generations to come.

# REFERENCES

Richmond T. J. (2017); Blockchain: 3 Books in 1; The Consice Guide to the Technology of Blockchain, Bitcoin, and Cryptocurrencies.

Scott Galloway; (2017); The Four; The Hidden DNA of Amazon, Apple, Facebook, and Google. Published by Penguin Audio

Max Tegmark; (2017); Life 3.0; Being Human in the Age of Artificial Intelligence. Audio book published by Random House Audio.

Gene Kim, Kevin Behr, George Spafford; (2014); The Phoenix Project; A Novel about IT, DevOps, and Helping Your Business Win 5th Anniversary Edition. (P)2015 Gene Kim, Kevin Behr, and George Spafford

Max Mckeown. (2014); The Innovation Book; How to Manage Ideas and Execution for Outstanding Results. (P)2015 Audible, Ltd.

Don Tapscott, Alex Tapscott; (2016); Blockchain Revolution; How the Technology Behind Bitcoin Is Changing Money, Business, and the World. Published by Brilliance Audio, all rights reserved. Recorded by arrangement with Porttolio, an imprint of Penguin Publishing Group, a division of Penguin Random House LLC.

Geoffrey A. Moore; (2012). Crossing the Chasm; Marketing and Selling Technology Projects to Mainstream Customers. Published by HarperCollins Publishers.

Malcolm Frank, Paul Roehrig, Ben Pring; (2017); What to Do When Machines Do Everything; How to Get Ahead in a World of AI, Algorithms, Bots, and Big Data. Cognizant Technology Solutions U.S. Corporation (P)2017 Audible, Inc.

Roger Schank; (2011); Teaching Minds; How Cognitive Science Can Save Our Schools. Published 2012 Redwood Audiobooks.

Tony Wagner; (2012); Creating Innovators; The Making of Young People Who Will Change the World. Published by Simon & Schuster

The Great Courses, Clinton O. Longenecker, Eric Sussman, Michael A. Roberto, Ryan Hamilton; (2015). Critical Business Skills for Success. Published by The Teaching Company, LLC.

Josh Kaufman; (2010); The Personal MBA: Master the Art of Business. Worldly Wisdom Ventures LLC.

Barry Z. Posner, James M. Kouzes; (2012); The Leadership Challenge; How to Make Extraordinary Things Happen in Organizations, Fifth Edition. Published by Gildan Media LLC.

Roo Rogers, Rachel Botsman; (2010); What's Mine Is Yours; The Rise of Collaborative Consumption. Published by Tantor

Bernard Marr; (2018); The 4th Industrial Revolution Is Here - Are You Ready? Retrieved from Forbes.com

Gideon Rose; (2016); The Fourth Industrial Revolution: A Davos Reader

A&E Television Networks; (October 29, 2009) Industrial Revolution; retrieved from https://www.history.com/topics/industrial-revolution/industrial-revolution; Access Date; 2019

John P. Rafferty; Industrial Revolution; Retrieved from https://www.britannica.com/event/ Industrial-Revolution

Marks, Robert B. (2002); The Origins of the Modern World: A Global and Ecological Narrative. Rowman and Littlefield; Lanham, MD.

James Chen; (2019); Industrial Revolution; Retrieved from https://www.investopedia.com

Uglow, Jenny. (2002); Five Friends Whose Curiosity Changed the World. Farrar, Straus, and Giroux; New York.

Matthew White; (2009); "The Industrial Revolution;" Georgian Britain; Retrieved from www.bl.uk/Georgian-Britain/articles

Strayer, Robert W. (2009); Ways of the World: A Brief Global History. "Revolutions of Industrialization." Bedford/St. Martin; Boston and New York.

More, Charles; (2000); Understanding the Industrial Revolution.

Laura Butler; (2018); The future of the workplace. Retrieved from hrtechnologist.com

Andreas Ortega; (2016); Employee Centricity: a Journey to the Centre of the Employee. Retrieved from glocalthinking.com

Vivian Maza; (2018); Employee-Centric Company Cultures. Retrieved from forbes.com

Ruck, K. & Welch, M. (2012) Valuing internal communication; management and employee perspectives. Public Relations Review, Vol. 38, Issue 2, pp. 294-302

Welch, M. & Jackson, P. R. (2007) Rethinking internal communication: a stakeholder approach. Corporate Communications: An International Journal, Vol. 12, Issue 2, pp. 177-198

Angela Stringfellow; (2018); Gig Economy. Retrieved from wonolo.com

Forbes; (2019); The Pros and Cons of the Gig Economy. Retrieved from forbes.com

John Frazer; (2019); How The Gig Economy Is Reshaping Careers For The Next Generation. Retrieved from

Angela Stringfellow; (2018); GIG ECONOMY. Retrieved from wonolo.com

Chelsea Levinson; (2018); Types of Competition in Economics. Retrieved from bizfluent.com

Connecting companies: Strategic partnerships for the digital age; A report from The Economist Intelligence Unit, commissioned by Telstra (2015)

Dirk Draheim, Gerald Weber eds. (2007) Trends in Enterprise Application Architecture: 2nd International Conference, TEAA 2006, Berlin, Germany, November 29 - December 1, 2006, Revised Selected Papers, p. 260

Feapo; (2017); the Federation of Enterprise Architecture Domains. Retrieved from wikipedia.org

Franz Emprechtinger; (2018); what is innovation culture? Retrieved from lead-innovation.com

Hitesh Bhasin; (2017); Types of Market. Retrieved from marketing91.com

Jonathan Auerbach; (2018); why the partnership is the business trend to watch. Retrieved from weforum.org

Lunar pages; (2018); The Power of Togetherness. Retrieved from lunarpages.com

Margaret Rouse; (20018); Project Management Body of Knowledge (PMBOK). Retrieved from whatis.techtarget.com

Margaret Rouse; (2015); Innovation culture. Retrieved from whatIs.com

Margaret Rouse; (2019); gig economy. Retrieved from techtarget.com

Martin; (2016); Introduction to the Gig Economy. Retrieved from cleverism.com

Narry Singh; (2016); the rise of digital partnerships. Retrieved from itproportal.com

Nick Ismail; (2018); Traditional companies need digital partners. Retrieved from information-age.com

Pearl Zhu; (2019); "Coopetition" as the digital Characteristic. Retrieved from futureofcio.blogspot.com

Phil McKinney; (2016); Innovation Culture: What does it mean and Why Does it Matter? Retrieved from philmckinney.com

Phillip Poarch; (2018); Coopetition: What Is It and How Can It Help Your Business Grow? Retrieved from channelfutures.com

Quick comics; (2016); the four types of market structure. Retrieved from quickonomics.com

Renodo; (2019); why innovation as a culture is important for your company? Retrieved from renodo.co.in

Rikke Dam; Teo Siang; (2019); Design Thinking: New Innovative Thinking for New Problems. Retrieved from interaction-design.org.

Ryoji Kimura; Martin Reeves; Kevin Whitaker; (2019); the new logic of competition. Retrieved from bcg.com

Sarah K. White; (2018); what is enterprise architecture? Retrieved from cio.com

Sylvie Laforet (2011), "A framework of organizational innovation and outcomes in SMEs", International Journal of Entrepreneurial Behavior & Research, Vol. 14, No. 4.

Telegraph; (2016); the rise of digital partnerships: fad or way forward? Retrieved from telegraph.co.uk

The Economist; (2019); Dawn of the digital partnership. Retrieved from connectedfuture.economist.com

Tien Nguyen; (2013); Building your partner ecosystem. Retrieved from openviewpartners.com

Laura Butler; (2018); The future of the workplace. Retrieved from hrtechnologist.com

Ginger Shimp; (2018); Digital Transformation's Impact On Business Processes and Work. Retrieved from digitalistmag.com

Sven Denecken; (2016); Effects of the Digital Transformation on Businesses. Retrieved from zdnet.com

Tsheets; (2019); What Is The Gig Economy-And How Does It Impact Employees? Retrieved from tsheets.com

Visual Paradigm; (2019); what is PMBOK in Project Management? Retrieved from visual-paradigm.com

Wiki; (2019); Business architecture. Retrieved from en.wikipedia.org

Will Kenton; (2018); Gig Economy. Retrieved from investopedia.com

Praba Shan; (2017); Data: The New Currency. Retrieved from digitalistmag.com

TEDx; (2017); is data the new currency? Retrieved from tedx.amsterdam

William D. Eggers, Rob Hamill, Abed Ali; (2013); Government's role in facilitating the exchange. Retrieved from www2.deloitte.com

Leonard, Kimberlee. (2018, October 25). The Role of Data in Business. Small Business - Chron.com. Retrieved from http://smallbusiness.chron.com/role-data-business-20405.html

Chris O'Connor; (2016); Data: the key measure of relevance in a digital revolution. Retrieved from ibm.com

Next Pathway; (2018); Big Data is Driving the Digital Revolution. Retrieved from nextpathway.com

Elena Alfaro; Juan Murillo; (2018); the three pillars of the digital revolution: data, talent, and innovation. Retrieved from bbva.com

Alan Marcus; (2015); Data and the fourth industrial revolution. Retrieved from weforum.org

Mikael Bohr; (2017); The Complex Role of Data in Today's Digital Revolution. Retrieved from teradata.com

WEF (2013) cybersecurity, data collection, privacy issue: Emerging issues in digital security. Geneva, Switzerland: World Economic Forum.

Angulo, J., Hübne, S. F., Wästlund, E., & Pulls, T. (2012). Towards usable privacy policy display and management. Information Management & Computer Security , 20 (1), 4-17.

Bélanger, F., and Crossler, R. (2011). Privacy in the digital age: a review of information privacy research in information systems",research in information systems. Management Information Systems Quarterly, 35 (3), 1017-1041

Ajit Patel; (2018); Cybersecurity and data privacy: what are you overlooking? Retrieved from itproportal.com

EPIC; (2019); Cybersecurity Privacy Practical Implications. Retrieved from epic.org

Sam Olens; (2019); Cybersecurity' and 'Privacy' Aren't the Same Things. Retrieved from law.com

Sanksshep Mahendra; (2018); A smarter way to manage cyber threats. Retrieved from medium.com

MacMillan C. Social media revolution and blurring of professional boundaries. Imprint.2013;60(3):44–46.

History Crunch; (2019); Impacts of the industrial revolution. Retrieved from historycrunch.com

Arun; (2018); 10 major effects of the industrial revolution. Retrieved from learnodo-newtonic.com

History Cooperative; (2018); Internet Business: A History. Retrieved from historycooperative.org

Preeti Seth; (2018); Before And After Internet: How Things Changed. Retrieved from blogs.systweak.com

K.A. Francis; (2019); How Has the Internet Impacted Businesses? Retrieved from smallbusiness.chron.com

Steve Ranger; (2018); Cybersecurity in an IoT and Mobile World. Retrieved from zdnet.com

123seminarsonly; (2019); Internet of Things. Retrieved from 123seminarsonly.com

Nils Herzberg; (2018); IoT And The 6 Categories Of Connected Things. Retrieved from digitalistmag.com

Knud Lueth; (2015); The 10 most popular Internet of Things applications right now. Retrieved from iot-analytics.com

Orbita; (2018); Four Pillars to the Internet of Everything. Retrieved blog.orbita.ai

Janice Abel; (2019); Extended Reality and Digital Twins – The Future Is Now. Retrieved from arcweb.com

Zach Capers; (2018); Digital twin: An example of related technologies converging. Retrieved from lab.getapp.com

Lanner; (2018); The Operational Process Digital Twin. Retrieved from lanner.com

Keith Shaw; Josh Fruhlinger; (2019); What is a digital twin? Retrieved from networkworld.com

John Spacey; (2016); 11 Examples of a Digital Twin. Retrieved from simplicable.com

Fitzgerald, M., Kruschwitz, N., Bonnet, D., Welch, M.: Embracing Digital Technology: A New Strategic Imperative. MIT Sloan Management Review, Research Report (2013)

Ross, J., Sebastian, I., Beath, C., Scantlebury, S., Mocker, M., Fonstad, N., Kagan, M., Moloney, K., Geraghty Krusel, S.: Designing Digital Organizations, vol. 46. MIT Center for

Research (2016) 3. Matt, C., Hess, T., Benlian, A.: Digital transformation strategies. Bus. Inf. Syst. Eng. 57(5),339–343 (2015) Digital Transformation 419

Ebert, C., Duarte, C.: Requirements engineering for the digital transformation: industry panel. In: Requirements Engineering Conference IEEE 24th International, pp. 4–5 (2016)

Westerman, G., Calméjane, C., Bonnet, D., Ferraris, P., McAfee, A.: Digital Transformation: A Roadmap for Billion-Dollar Organizations, pp. 1–68. MIT Sloan Management, MIT Center for Digital Business and Capgemini Consulting (2011)

Bharadwaj, A.: A resource-based perspective on information technology capability and firm performance: an empirical investigation. MIS Q. 24(1), 169–196 (2000)

Sebastian, I., Ross, J., Beath, C., Mocker, M., Moloney, K., Fonstad, N.: How Big Old Companies Navigate Digital Transformation. MIS Quarterly Executive (2017)

Zinder, E., Yunatova, I.: Synergy for digital transformation: person's multiple roles and subject domains integration. In: Digital Transformation and Global Society, pp. 155–168 (2016)

Hess, T., Matt, C., Benlian, A., Wiesböck, F.: Options for formulating a digital transformation strategy. MIS Q. Executive 15(2), 123–139 (2016)

Carcary, M., Doherty, E., Conway, G.: A dynamic capability approach to digital transformation–a focus on key foundational themes. In: 10th European Conference on Information Systems Management. Academic Conferences and publishing limited, pp. 20–28 (2016)

Digital Transformation: A Roadmap for Billion-Dollar Organizations, by Capgemini and MIT Center for Digital Business

The 2014 State of Digital Transformation, by Altimeter

The Nine Elements of Digital Transformation, by George Westerman, Didier Bonnet and Andrew McAfee

Westerman, G., Bonnet, D., & McAfee, A. (2012). The advantages of digital maturity. MIT Sloan Management Review. Retrieved July 21, 2017, from http://sloanreview.mit.edu/article/the-advantages-of-digital-maturity/ [Google Scholar]

JD Meier; (2016); Digital Transformation Defined. Retrieved from blogs.msdn.microsoft.com

Anderson, J. & Lanzolla, G., 2010. The Digital Revolution is Over: Long Live the Digital Revolution! Business Strategy Review, 21(1), pp.74–77.

Schwab, K., 2016. The Fourth Industrial Revolution, World Economic Forum.

Atkinson, R., 2005. Prospering in an era of economic transformation. Economic Development Journal, Summer, pp.33–38.

Lasi, H. et al., 2014. Industry 4.0. Business and Information Systems Engineering, 6(4), pp.239–242.

Gray, P. et al., 2013. Realizing Strategic Value Through Center-Edge Digital Transformation in Consumer-Centric Industries. MIS Quarterly Executive, 12(1), pp.115–117.

Basole, R.C., 2016. Accelerating Digital Transformation Visual Insights from the API Ecosystem. IT Pro (December), pp.20–25.

Belk, R.W., 2013. Extended Self in a Digital World. Journal of Consumer Research, 40(3), pp.477–500.

Austin, R.D. & Upton, D.M., 2016. Leading in the Age of Super-Transparency. MIT Sloan Management Review, 57(2), pp.25–32.

Hess, T. et al., 2016. Options for Formulating a Digital Transformation Strategy. MIS Quarterly Executive, 15(2), pp.123–139.

Salesforce; (2019); Examples of Digital Transformation. Retrieved from salesforce.com

Christopher Penn; (2018); Transforming People, Process, and Technology. Retrieved christopherspenn.com

Willem Sundblad; (2018); How Manufacturers Can Get IT and OT to Work Together. Retrieved from forbes.com

John Koon; (2018); The impact of OT/IT convergence on manufacturing. Retrieved from enterpriseiotinsights.com

Dave Gorin; (2018); The connected factory. Retrieved from traccsolution.com

Glen Hartman; (2013); Three key steps towards a customer-focused digital transformation. Retrieved from econsultancy.com

Martin Roll; (2017); Customer-centric and Consumer-driven Brands. Retrieved from martinroll.com

Answering the Digital and Analytics Talent Gap: The New "Trilinguals" at www.atkearney.com. April 2017

Jill Rowley; (2017); A Modern Marketing Map for Customer-centric Success. Retrieved from digitalmarketinginstitute.com

Anubhav Rawat; (2018); Achieving Customer-centricity through Digital Transformation. Retrieved from wns.com

Julio Hernandez; (2019); The four fundamentals of customer-centric digital transformation    Retrieved from advisory.kpmg.us

"Gig Economy," Investopedia (24 May 2018), online.

The Gig Economy: Achieving Financial Wellness with Confidence", BMO Wealth Management (30 July 2018), online; see also Elaine Pofeldt, "Why Older Workers Are Embracing the Gig Economy," Forbes (30 August 2017), online.

Sunil Johal, Sara Ditta & Noah Zon, "Taxi and Limousine Regulations and Service Review – Emerging Issues in the Taxi and Limousine Industry," Mowat Centre (22 October 2015) at 7-8, online.

City of Ottawa - Taxi and Limousine Regulations and Services Review: Customer Experience", Core Strategies (14 October 2015), at 8; Ashley

Csanady, "If the Uber debate is really about safety, why are women's voices being sidelined?", National Post (26 April 2016), online.

Gene Demby, "Apps Make Googly Eyes At Riders Tired Of Being Snubbed By Cabbies," National Public Radio (21 October 2014), online.

Steve Hawk, "What an economist, learned by driving for Uber," Quartz (5 March 2018), online.

Sherissa Rajah; Shane Todd; (2019); The Canadian Gig Economy: Embracing the Future of Work. Retrieved from fasken.com

Abdullahi Muhammed; (2019); 5 Important Stats About The Gig Economy To Know In 2019. Retrieved from forbes.com

AJ O'Connell; (2019); 6 Trends To Watch Within the Gig Economy for 2019. Retrieved from blog.steadyapp.com

The Conversation; (2019); Work in the 'gig economy': one-night stand or a meaningful relationship? Retrieved from theconversation.com

Alberto Dominguez; (2018); 8 advantages of digitalization of business. Retrieved from ehorus.com

Shane Kos; (2017); Has your business gone digital? Retrieved from aureon.com

Marzo; (2018); How can your company benefit from digital transformation? Retrieved from mydatascope.com

Idexcel Technologies; (2018); Top 5 Key Benefits of Digital Transformation. Retrieved from idexcel.com

# INDEX

Where to study Online as discussed chapter 18, 20 and 21

http://www.mit.edu/

https://online.hbs.edu/

https://www.stanford.edu/

https://www.yale.edu/

https://hbr.org/

https://www.coursera.org/

https://www.edx.org/

lynda.com

https://www.udemy.com/

https://www.udacity.com/

https://su.org/

https://www.reuters.com

https://www.cisco.com/

Thank you for taking the time to read this book, I hope you found it useful. My final word to you, we need enough people working together to ensure that technology and digitization is put of the best possible use for humanity, and to subside and push away all evil and harmful usage; and for you to keep up with the pace at which technology is transforming and revolutionizing the world, embrace technology in the fullness of its potentials in everything you do!!!

---

# THANK YOU.

---